Health Risks from Dioxin and Related Compounds

Evaluation of the EPA Reassessment

Committee on EPA's Exposure and Human Health
Reassessment of TCDD and Related Compounds

Board on Environmental Studies and Toxicology
Division on Earth and Life Studies

NATIONAL RESEARCH COUNCIL
OF THE NATIONAL ACADEMIES

THE NATIONAL ACADEMIES PRESS
Washington, DC
www.nap.edu

THE NATIONAL ACADEMIES PRESS 500 Fifth Street, N.W. Washington, DC 20001

NOTICE: The project that is the subject of this report was approved by the Governing Board of the National Research Council, whose members are drawn from the councils of the National Academy of Sciences, the National Academy of Engineering, and the Institute of Medicine. The members of the committee responsible for the report were chosen for their special competences and with regard for appropriate balance.

This project was supported by Contract No. 68-C-03-081 between the National Academy of Sciences and the U.S. Environmental Protection Agency. Any opinions, findings, conclusions, or recommendations expressed in this publication are those of the author(s) and do not necessarily reflect the view of the organizations or agencies that provided support for this project.

International Standard Book Number-10: 0-309-10258-8 (Book)
International Standard Book Number-13: 978-0-309-10258-2 (Book)
International Standard Book Number-10: 0-309-66273-7 (PDF)
International Standard Book Number-13: 978-0-309-66273-4 (PDF)

Library of Congress Control Number 2006933608

Additional copies of this report are available from

The National Academies Press
500 Fifth Street, NW
Box 285
Washington, DC 20055

800-624-6242
202-334-3313 (in the Washington metropolitan area)
http://www.nap.edu

Printed in the United States of America

THE NATIONAL ACADEMIES

Advisers to the Nation on Science, Engineering, and Medicine

The **National Academy of Sciences** is a private, nonprofit, self-perpetuating society of distinguished scholars engaged in scientific and engineering research, dedicated to the furtherance of science and technology and to their use for the general welfare. Upon the authority of the charter granted to it by the Congress in 1863, the Academy has a mandate that requires it to advise the federal government on scientific and technical matters. Dr. Ralph J. Cicerone is president of the National Academy of Sciences.

The **National Academy of Engineering** was established in 1964, under the charter of the National Academy of Sciences, as a parallel organization of outstanding engineers. It is autonomous in its administration and in the selection of its members, sharing with the National Academy of Sciences the responsibility for advising the federal government. The National Academy of Engineering also sponsors engineering programs aimed at meeting national needs, encourages education and research, and recognizes the superior achievements of engineers. Dr. Wm. A. Wulf is president of the National Academy of Engineering.

The **Institute of Medicine** was established in 1970 by the National Academy of Sciences to secure the services of eminent members of appropriate professions in the examination of policy matters pertaining to the health of the public. The Institute acts under the responsibility given to the National Academy of Sciences by its congressional charter to be an adviser to the federal government and, upon its own initiative, to identify issues of medical care, research, and education. Dr. Harvey V. Fineberg is president of the Institute of Medicine.

The **National Research Council** was organized by the National Academy of Sciences in 1916 to associate the broad community of science and technology with the Academy's purposes of furthering knowledge and advising the federal government. Functioning in accordance with general policies determined by the Academy, the Council has become the principal operating agency of both the National Academy of Sciences and the National Academy of Engineering in providing services to the government, the public, and the scientific and engineering communities. The Council is administered jointly by both Academies and the Institute of Medicine. Dr. Ralph J. Cicerone and Dr. Wm. A. Wulf are chair and vice chair, respectively, of the National Research Council.

www.national-academies.org

COMMITTEE ON EPA'S EXPOSURE AND HUMAN HEALTH REASSESSMENT OF TCDD AND RELATED COMPOUNDS

Members

David L. Eaton (*Chair*), University of Washington, Seattle
Dennis M. Bier, Baylor College of Medicine, Houston, TX
Joshua T. Cohen, Tufts New England Medical Center, Boston, MA
Michael S. Denison, University of California, Davis
Richard T. Di Giulio, Duke University, Durham, NC
Norbert E. Kaminski, Michigan State University, East Lansing
Nancy K. Kim, New York State Department of Health, Troy
Antoine Keng Djien Liem, European Food Safety Authority, Parma, Italy
Thomas E. McKone, Lawrence Berkeley National Laboratory and School of Public Health, University of California, Berkeley
Malcolm C. Pike, University of Southern California, Los Angeles
Alvaro Puga, University of Cincinnati Medical Center, Cincinnati, OH
Andrew G. Renwick, University of Southampton (emeritus), Southampton, UK
David A. Savitz, Mount Sinai School of Medicine, New York, NY
Allen E. Silverstone, SUNY–Upstate Medical University, Syracuse, NY
Paul F. Terranova, University of Kansas Medical Center, Kansas City
Kimberly M. Thompson, Massachusetts Institute of Technology, Cambridge
Gary M. Williams, New York Medical College, Valhalla
Yiliang Zhu, University of South Florida, Tampa

Staff

Suzanne van Drunick, Project Director
Kulbir Bakshi, Senior Program Officer for Toxicology
Ruth Crossgrove, Senior Editor
Jean Hampton, Senior Fellow
Cay Butler, Editor
Mirsada Karalic-Loncarevic, Research Associate
Bryan P. Shipley, Research Associate
Liza R. Hamilton, Senior Program Assistant
Alexandra Stupple, Senior Editorial Assistant
Sammy Bardley, Librarian

Sponsors

U.S. Environmental Protection Agency
U.S. Department of Agriculture
U.S. Department of Health and Human Services

OTHER REPORTS OF THE BOARD ON ENVIRONMENTAL STUDIES AND TOXICOLOGY

Assessing the Human Health Risks of Trichloroethylene: Key Scientific Issues (2006)
New Source Review for Stationary Sources of Air Pollution (2006)
Human Biomonitoring for Environmental Chemicals (2006)
Fluoride in Drinking Water: A Scientific Review of EPA's Standards (2006)
State and Federal Standards for Mobile-Source Emissions (2006)
Superfund and Mining Megasites—Lessons from the Coeur d'Alene River Basin (2005)
Health Implications of Perchlorate Ingestion (2005)
Air Quality Management in the United States (2004)
Endangered and Threatened Species of the Platte River (2004)
Atlantic Salmon in Maine (2004)
Endangered and Threatened Fishes in the Klamath River Basin (2004)
Cumulative Environmental Effects of Alaska North Slope Oil and Gas Development (2003)
Estimating the Public Health Benefits of Proposed Air Pollution Regulations (2002)
Biosolids Applied to Land: Advancing Standards and Practices (2002)
The Airliner Cabin Environment and Health of Passengers and Crew (2002)
Arsenic in Drinking Water: 2001 Update (2001)
Evaluating Vehicle Emissions Inspection and Maintenance Programs (2001)
Compensating for Wetland Losses Under the Clean Water Act (2001)
A Risk-Management Strategy for PCB-Contaminated Sediments (2001)
Acute Exposure Guideline Levels for Selected Airborne Chemicals (4 volumes, 2000-2004)
Toxicological Effects of Methylmercury (2000)
Strengthening Science at the U.S. Environmental Protection Agency (2000)
Scientific Frontiers in Developmental Toxicology and Risk Assessment (2000)
Ecological Indicators for the Nation (2000)
Waste Incineration and Public Health (1999)
Hormonally Active Agents in the Environment (1999)
Research Priorities for Airborne Particulate Matter (4 volumes, 1998-2004)
The National Research Council's Committee on Toxicology: The First 50 Years (1997)
Carcinogens and Anticarcinogens in the Human Diet (1996)
Upstream: Salmon and Society in the Pacific Northwest (1996)

Science and the Endangered Species Act (1995)
Wetlands: Characteristics and Boundaries (1995)
Biologic Markers (5 volumes, 1989-1995)
Review of EPA's Environmental Monitoring and Assessment Program (3 volumes, 1994-1995)
Science and Judgment in Risk Assessment (1994)
Pesticides in the Diets of Infants and Children (1993)
Dolphins and the Tuna Industry (1992)
Science and the National Parks (1992)
Human Exposure Assessment for Airborne Pollutants (1991)
Rethinking the Ozone Problem in Urban and Regional Air Pollution (1991)
Decline of the Sea Turtles (1990)

Copies of these reports may be ordered from the National Academies Press
(800) 624-6242 or (202) 334-3313
www.nap.edu

Acknowledgments

We are appreciative of the generous support provided by the U.S. Environmental Protection Agency and are especially grateful for the outstanding assistance provided by Dr. William Farland. We are also grateful to Lisa Matthews, EPA's program manager, and for Dr. Richard Canady's assistance in facilitating invited speakers from the federal agencies.

Many people assisted the committee and National Research Council staff in creating this report. We are grateful for the information and support provided by the following:

Lesa L. Aylward, Summit Toxicology, L.L.P.
P. Michael Bolger, U.S. Food and Drug Administration
Gail Charnley, HealthRisk Strategies (on behalf of the Food Industry Dioxin Working Group)
Richard W. Clapp, Boston University School of Public Health
Edmund A. C. Crouch, Cambridge Environmental Inc.
Christopher T. De Rosa, Agency for Toxic Substances and Disease Registry
Michael J. DeVito, U.S. Environmental Protection Agency
David W. Gaylor, Gaylor and Associates, LLC
David P. Goldman, U.S. Department of Agriculture
C.T. 'Kip' Howlett, Consultant
Russell E. Keenan, AMEC Earth & Environmental Inc.
Larry L. Needham, Centers for Disease Control and Prevention
Christopher J. Portier, National Institute of Environmental Health Sciences
Susan Schober, Centers for Disease Control and Prevention
Jay B. Silkworth, General Electric Company
Nigel Walker, National Institute of Environmental Health Sciences

The committee's work also benefited from written and oral testimony submitted by the public, whose participation is much appreciated.

Acknowledgment of Review Participants

This report has been reviewed in draft form by individuals chosen for their diverse perspectives and technical expertise, in accordance with procedures approved by the National Research Council's Report Review Committee. The purpose of this independent review is to provide candid and critical comments that will assist the institution in making its published report as sound as possible and to ensure that the report meets institutional standards for objectivity, evidence, and responsiveness to the study charge. The review comments and draft manuscript remain confidential to protect the integrity of the deliberative process. We wish to thank the following individuals for their review of this report:

Melvin Andersen, CIIT Centers for Health Research
John Doull, University of Kansas Medical Center
David Gaylor, Gaylor & Associates
Michael Holsapple, ILSI Health and Environmental Sciences
Daniel Krewski, University of Ottawa
Philip Landrigan, Mount Sinai School of Medicine
John A. Moore, Hollyhouse, Inc.
Stephen S. Olin, ILSI Research Foundation/Risk Science
Richard Peterson, School of Pharmacy, Harvard School of Public Health
Louise Ryan, Harvard School of Public Health
Steven Safe, Texas A&M University
Glenn Sipes, University of Arizona
Martin Van den Berg, Utrecht University

Noel Weiss, University of Washington
Lauren Zeise, California Environmental Protection Agency

Although the reviewers listed above have provided many constructive comments and suggestions, they were not asked to endorse the conclusions or recommendations, nor did they see the final draft of the report before its release. The review of this report was overseen by William Halperin and John Bailar. Appointed by the National Research Council, they were responsible for making certain that an independent examination of this report was carried out in accordance with institutional procedures and that all review comments were carefully considered. Responsibility for the final content of this report rests entirely with the authoring committee and the institution.

Preface

2,3,7,8-Tetrachlorodibenzo-*p*-dioxin (TCDD), also called dioxin, is among the most toxic anthropogenic substance ever identified. TCDD and a number of similar polychlorinated dioxins, dibenzofurans, and coplanar polychlorinated biphenyls (dioxin-like compounds [DLCs]) have been the subject of intense scientific research and frequently controversial environmental and health policies. Animal studies have demonstrated potent effects of TCDD, other dioxins, and many DLCs on tumor development, birth defects, reproductive abnormalities, immune dysfunction, dermatological disorders, and a plethora of other adverse effects. Because of their persistence in the environment and their bioaccumulative potential, TCDD, other dioxins, and DLCs are now ubiquitous environmental pollutants and are detected at low concentrations in virtually all organisms at higher trophic levels in the food chain, including humans. Inadvertent exposures of humans through industrial accidents, occupational exposures to commercial compounds (primarily phenoxyacid herbicides), and through dietary pathways have led to a wide range of body burdens of TCDD, other dioxins, and DLCs, and numerous epidemiological studies have attempted to relate exposures to a variety of adverse effects in humans.

Because of substantial policy and economic implications associated with the regulation of TCDD, other dioxins, and DLCs in the environment, the U.S. Environmental Protection Agency (EPA) began in the mid-1980s to invest enormous efforts in risk assessment of these compounds. Many scientists in the dioxin research community participated in writing numerous review chapters on various aspects of dioxin toxicology, chemistry, and environmental fate. In September 1992, initial drafts of all background chapters of the EPA assessment underwent extensive peer review, followed by extensive revision and additional review of some chapters. In September 1994, all the chapters, plus the first draft of a summary "risk characterization" chapter, were subjected to more peer review and public comment. In 1997 and 1998, additional modifications of the compiled information led to the development of an "Integrated Summary and Risk Characterization" document. This document, as well as additional information on toxic equivalency of DLCs, was revised and subsequently reviewed by EPA's Science Advisory Board (SAB) in November 2000. Recognizing the broad policy implications of the dioxin reassessment, an Interagency Working Group (IWG), consisting of representatives of seven federal agencies, was established in 2000 to foster information sharing, develop a common language for dioxin science and science policy across governmental agencies and programs, identify gaps and needs in dioxin risk assessment, and coordinate risk management strategies. The IWG has provided input to EPA on the draft dioxin reassessment and has been coordinating risk management issues on TCDD and other dioxins for the federal government since its inception. After further revisions in response to SAB and other public comments, in December 2003, EPA released a preliminary draft document titled *Exposure and Human Health Reassessment of 2,3,7,8-Tetrachlorodibenzo-p-Dioxin (TCDD) and Related Compounds*, referred to in this report as the Reassessment.

In the summer of 2004, EPA requested the National Research Council (NRC) to create "an expert committee to review EPA's draft reassessment of the risks of dioxin and dioxin-like compounds." In response, the NRC appointed the Committee on EPA's Exposure and Human Health Reassessment of TCDD and Related Compounds, which was charged, to the extent possible, to review "EPA's modeling assumptions, including those associated with dose-response curve and points-of-departure dose ranges and associated likelihood estimates for identified human health outcomes; EPA's quantitative uncertainty analysis; EPA's selection of studies as a basis for its assessments and gaps in scientific knowledge." The charge also requested that the committee address two specific points of controversy: (1) the scientific evidence for classifying dioxin as a human carcinogen, and (2) the validity of the nonthreshold linear dose-response model and the cancer slope factor calculated by EPA through the use of this model. The committee was also asked to comment on the usefulness of toxic equivalency factors (TEFs) and the uncertainties associated with their use in risk assessment of complex mixtures. Finally, the committee was also asked to review the uncertainty associated with the Reassessment's approach to the analysis of food sampling and human dietary intake data.

The entire Reassessment consists of three parts totaling more than 1,800 pages of scientific review. Part I contains several volumes of a previous scientific review of information relating to sources and exposures to TCDD and other dioxins in the environment, and Part II contains detailed reviews of scientific information on the health effects of TCDD, other dioxins, and DLCs. The information in Parts I and II were provided to the committee as background, with the recognition that many chapters in these two volumes have not been updated for several years. The committee was asked to focus its review on Part III of the Reassessment, which represents an "integrated summary and risk characterization for TCDD and related compounds."

The committee held five meetings between November 22, 2004, and July 7, 2005. The first three meetings provided opportunity for public input. The committee heard from scientists from the IWG, EPA, Food and Drug Administration, Department of Agriculture, Agency for Toxic Substance and Disease Registry, National Center for Health Statistics, and National Toxicology Program and from representatives from academia, environmental organizations, and the regulated community. The committee was provided with written testimony and new scientific papers that have appeared since 2003 (and thus were not available for consideration by EPA in the Reassessment).

It is important to recognize what the committee did not consider to be part of its charge. Although the committee made every effort to consider critical new studies that have appeared since the last revision of Part III of the Reassessment, it did not conduct an exhaustive and detailed review of all scientific information published on TCDD and other dioxins since 2003, and any information that became available to the public after the date of the committee's last meeting (July 7, 2005) was not considered. The committee did not attempt to "redo" the risk assessment—rather, it tried to provide constructive comments in areas in which the scientific approaches or justifications were thought to need improvement, the expectation being that EPA might need to reconsider and revise its approaches and documentation accordingly.

The final recommendations of the committee are offered to EPA with the recognition and appreciation of the enormous amount of time and effort that has been committed to the execution of this Reassessment for nearly 14 years. Although many of the comments are, not surprisingly, critical of certain aspects or approaches taken by EPA, the committee was impressed overall with the tremendous dedication and hard work that has gone into the creation of the Reassessment. The committee hopes the report will be of value in assisting EPA to make final changes to Part III that will allow the timely release of a scientifically defensible document. The committee further hopes that this review will help to guide all federal agencies in making rational and defensible health and environmental policies that adequately protect human health and the environment from the adverse effects of TCDD, other dioxins, and DLCs in the environment.

The Committee on EPA's Exposure and Human Health Reassessment of TCDD and Related Compounds was aided immensely by a number of individuals. The committee, and especially the chair, would like to thank the NRC study director Suzanne van Drunick for her tireless effort and good humor in directing this project under substantial time constraints. We also appreciate the organizational skills of Liza Hamilton for ensuring that our meetings and travel arrangements went smoothly, and other NRC staff, including Bryan Shipley for his technical assistance, Ruth Crossgrove and Cay Butler for their editorial assistance, Mirsada Karalic-Loncarevic for her reference assistance, and Alexandra Stupple for her production assistance. The committee is also grateful to Kulbir Bakshi, senior program officer; James Reisa, director of the Board on Environmental Studies and Toxicology; and Thomas Burke, professor and associate chair, Johns Hopkins University, for their oversight of the study; and to Ann Yaktine, Food and Nutrition Board, Institute of Medicine, for her contribution. I would like to thank all the committee members for their hard work and their dedication to ensuring that the report stands up to the basic charge that we "ensure that the risk estimates ... are scientifically robust." I, the NRC staff, and the committee are indebted to a number of individuals who presented background information, both orally and in writing, that made the committee's understanding of the issues more complete. Thanks especially to Richard Canady, IWG on dioxin, for his assistance in helping to locate speakers and important background documents and to William Farland for his outstanding assistance.

David L. Eaton, *Chair*
Committee on EPA's Exposure and Human Health
Reassessment of TCDD and Related Compounds

Abbreviations

2-AAF: 2-acetylaminofluorene
AHF: altered hepatocelluar foci
AHR: aromatic hydrocarbon receptor
Ahr$^{-/-}$: AHR null
AIC: Akaike's information criterion
Anti-SRBC: anti-sheep red blood cell
ARNT: AHR nuclear translocator protein
ATSDR: Agency for Toxic Substances and Disease Registry
AUC: area under the curve
BMD: benchmark dose
BMDL: benchmark dose low
BMR: benchmark response
CB: chlorobiphenyl
CI: confidence intervals
CL: volume of blood cleared per unit time
CLB: cumulative lipid burden
COX: cyclooxygenase
COX-2: cyclooxygenase-2
CSF: cancer slope factor
CYP1A: cytochrome P450A1 protein
CYP1A1: cytochrome P4501A1 protein
CYP1A2: cytochrome P4501A2 protein
CYP1B1: cytochrome P4501B1 protein
DHHS: U.S. Department of Health and Human Services
DIM: diindolymethane

DLCs: dioxin-like compounds
DOD: U.S Department of Defense
DF: dioxins and furons
DFP: dioxins, furons, and PCBs
ED: effective dose
EGFR: epidermal growth factor receptor
EPA: U.S. Environmental Protection Agency
ER: estrogen receptor
FAO: Food and Agriculture Organization of the United Nations
FDA: U.S. Food and Drug Administration
FSH: follicle-stimulating hormone
GGT: γ-glutamyl transpeptidase
GnRH: gonadotropin-releasing hormone
HAH: halogenated aromatic hydrocarbon
hCG: human chorionic gonadotropin
HpCDD: heptachlorodibenzo-*p*-dioxin
HepCB: heptachlorobiphenyl
HxCDD: hexachlorodibenzo-*p*-dioxin
HxCDF: hexachlorodibenzofuran
I3C: indole-3-carbinol
IARC: International Agency for Research on Cancer
ICZ: indolo-[3,2b]-carbazole
IOM: Institute of Medicine
IPCS: International Program of Chemical Safety
IWG: Interagency Working Group
JECFA: Joint Expert Committee on Food Additives
LABB: lifetime average body burden
LD: lethal dose
LED: lowest effective dose
LH: lutenizing hormone
LOAEL: lowest-observed-adverse-effect level
LOD: limit of detection
6-MCDF: 6-methyl-1,3,8-trichlorodibenzofuran
MOE: margin of exposure
mRNA: messenger ribonucleic acid
NAS: National Academy of Sciences
NCEA: National Center for Environmental Assessment
NIEHS: National Institute of Environmental Health Sciences
NIH: National Institutes of Health
NIOSH: National Institute for Occupational Safety and Health
NOAEL: no-observed-adverse-effect level
NOEL: no-observed-effect level
NRC: National Research Council

NTP: National Toxicology Program
OCDF: octachlorodibenzofuran
OCDD: octachlorodibenzo-*p*-dioxin
PA: plasminogen activator
PAH: polycyclic aromatic hydrocarbon
PAI-1: plasminogen activator inhibitor-1
PBDD: polybrominated dibenzo-*p*-dioxin
PBDF: polybrominated dibenzofuran
PBPK: physiologically based pharmacokinetics
PCB: polychlorinated biphenyl
PCDD: polychlorinated dibenzo-*p*-dioxin
PCDF: polychlorinated dibenzofuran
PeCB: pentachlorobiphenyl
PeCDD: pentachlorodibenzo-*p*-dioxin
PeCDF: pentachlorodibenzofuran
PK: pharmacokinetics
POD: point of departure
PPAR: peroxisome proliferator activated receptor
ppt: parts per trillion
PR: progesterone receptor
QF: quality of fit
REP: relative potency
RfD: reference dose
RR: rate ratio
SAB: Science Advisory Board
SCF: Scientific Committee on Food
SD: standard deviation
SE: standard error
SMR: standardized mortality (morbidity) ratio
T3: triiodothyronine
T4: thyroxine
TCB: 2,2′,5,5′-tetrachlorobiphenyl
TCDD: 2,3,7,8-tetrachlorodibenzo-*p*-dioxin
TCDF: 2,3,7,8-tetrachlorodibenzo furon
TEF: toxic equivalency factor
TEQ: toxic equivalent quotient
tPA: tissue plasminogen activator
2,4,5-T: 2,4,5-trichlorophenoxyacetic acid
TSH: thyroid-stimulating hormone
UED: upper effective dose
USDA: U.S. Department of Agriculture
WHO: World Health Organization

Contents

BOXES

Health Risks from Dioxin and Related Compounds

Evaluation of the EPA Reassessment

Public Summary

HEALTH RISKS FROM TCDD, OTHER DIOXINS, AND DIOXIN-LIKE COMPOUNDS

Evaluation of the EPA Reassessment

Dioxins and dioxin-like compounds (DLCs) are released into the environment from several sources, including combustion, metal processing, and chemical manufacturing and processing. The most toxic of these compounds is TCDD, often simply called dioxin. Many other types of dioxins, other than TCDD, and DLCs share most, if not all, of the toxic characteristics of TCDD. In the past, occupational exposures to TCDD, other dioxins, and DLCs occurred in a variety of industries, especially those involved in the manufacture of trichlorophenol (used to make certain herbicides) and PCBs. (PCBs contain some forms that are dioxin-like and, when heated to high temperatures, may also be contaminated with dibenzofurans, which are also dioxin-like.) Much of the knowledge about the health effects of TCDD, other dioxins, and DLCs in humans comes from studies of relatively highly exposed workplace populations. Widespread use of certain herbicides containing TCDD, other dioxins, and DLCs, as well as some types of industrial emissions, resulted in local and global contamination of air, soil, and water with trace levels of these compounds. These trace levels built up in the food chain because TCDD, other dioxins, and DLCs do not readily degrade. Instead, they persist in the environment and accumulate in the tissues of animals. The general public is exposed to TCDD, other dioxins, and DLCs primarily by eating such foods as beef, dairy products, pork, fish, and shellfish.

The health effects of exposures to relatively high levels of dioxin became widely publicized due to the use of the herbicide called Agent Orange in the Vietnam War. Agent Orange contained small amounts of TCDD as a contaminant. Studies suggest that veterans and workers exposed occupationally to TCDD, other dioxins, and DLCs experience an increased risk of developing a potentially disfiguring skin lesion (called chloracne), liver disease, and possibly cancer and diabetes.

Fortunately, background exposures for most people are typically much lower than those seen in either Vietnam veterans or occupationally exposed workers. The potential adverse effects of TCDD, other dioxins, and DLCs from long-term, low-level exposures to the general public are not directly observable and remain controversial. One major controversy is the issue of estimating risks at doses below the range of existing reliable data. Another controversy is the issue of appropriately assessing the toxicity of various mixtures of these compounds in the environment.

In 2004, the U.S. Environmental Protection Agency (EPA), asked the National Research Council (NRC) of the National Academies to review its 2003 draft document titled *Exposure and Human Health Reassessment of 2,3,7,8-Tetrachlorodibenzo-p-Dioxin (TCDD) and Related Compounds* (the Reassessment). This NRC report describes the Reassessment as very comprehensive in its review and analysis of the extensive scientific literature on TCDD, other dioxins, and DLCs. However, the NRC report finds substantial room for improvement in the quantitative approaches used by EPA to characterize risks. In particular, the committee recommends that EPA more thoroughly justify and communicate its approaches to dose-response modeling for health effects and make its criteria for selection of key data sets more transparent. EPA should also improve how it handles and communicates the substantial uncertainty that surrounds its various estimates of health risks from low-level exposures to TCDD, other dioxins, and DLCs. This NRC report provides a critical review of EPA's Reassessment, but the report is not a risk assessment and does not recommend exposure levels for TCDD, other dioxins, or DLCs for regulatory consideration. Rather the NRC report provides guidance to EPA on how the agency could improve the scientific robustness and clarity of the Reassessment for its ultimate use in risk management of TCDD, other dioxins, and DLCs in the environment by federal, state, and local regulatory agencies.

Assessing Human Exposure to TCDD, Other Dioxins, and DLCs

People worldwide are exposed to background levels of TCDD, other dioxins, and DLCs. Background exposures include those from the commercial food supply, air, water, and soil. EPA's 2003 draft Reassessment does not identify many specific direct sources of human exposures to relatively high levels of TCDD, other dioxins, or DLCs. EPA estimated background concentrations based on studies conducted at various locations in North America. Those studies examined a small number of locations and, hence, may not fully characterize national variability. EPA derived its estimates of TCDD, other dioxins, and DLCs in food from statistically based national surveys, nationwide-sampling networks, food fat concentrations, and environmental samples of air, water, soil, and food.

According to recent estimates, background concentrations of TCDD, other dioxins, and DLCs continue to decline. EPA's estimates of releases of these compounds to air, water, and land from reasonably quantifiable sources in 2000 showed a decrease of 89% from its 1987 estimates. At least one U.S. study determined that meat contains lower levels of TCDD, other dioxins, and DLCs than samples from the 1950s through the 1970s. An ongoing national study by the U.S. Department of Agriculture of the concentrations of TCDD, other dioxins, and DLCs in beef, pork, and poultry should allow for a time trend analysis of food concentrations.

To assess the total magnitude of emissions of TCDD, other dioxins, and DLCs, EPA used a "bottom-up" approach that attempted to identify all emission-source categories (such as combustion, metal processing, and chemical manufacturing and processing) and then estimated the magnitude of emissions for each category. The committee concludes that a "top-down" approach would also provide useful information and could give rise to significantly different estimates of the historical levels of emissions of TCDD, other dioxins, and DLCs. A top-down approach would account for measured levels in humans and the environment and consider the emission sources required to account for these levels.

The committee also recommends that EPA set up an active database of *typical* concentrations for TCDD, other dioxins, and DLCs present in food. This database should be based on a collection of all available data and updated on a regular basis with new data as they are published in the peer-reviewed literature.

Cancer Risk and TCDD, Other Dioxins, and DLCs

The EPA Reassessment revisits EPA's classification of TCDD, other dioxins, and DLCs on their potential to cause cancer in humans. In 1985, EPA classified TCDD as a "probable human carcinogen" based on the data available and EPA's classification criteria in place at the time. The Reassessment, which revisited this issue given the current evidence and a different draft classification scheme, characterized TCDD as "carcinogenic to humans." In 2005, after completion of the Reassessment, EPA further revised its cancer guidelines. In its charge, the NRC committee was specifically asked to address "the scientific evidence for classifying TCDD as a human carcinogen."[1] Referring to the definitions of chemical carcinogens in the EPA's current cancer guidelines, the NRC committee was split on whether the evidence from available studies met *all* the criteria necessary for definitive classification of TCDD as "carcinogenic to humans," although the committee unanimously agreed on a classification for TCDD of at least "likely to be carcinogenic to humans." The committee believed that the *public health* implications of the two terms appeared identical and for this reason did not belabor the issue of classification. The committee concluded that because the definition of "carcinogenic to humans" changed somewhat from previous EPA guidelines and after submission of the Reassessment, EPA should reevaluate its 2003 conclusion based on the criteria set out in its 2005 cancer guidelines.

The committee agrees with EPA in classifying other dioxins and DLCs as "likely to be carcinogenic to humans." However, because mixtures of DLCs and other dioxins may include TCDD, EPA should reconsider its classification of such mixtures as "likely to be carcinogenic to humans" if it continues to classify TCDD as "carcinogenic to humans."

Estimating Cancer Risks at Very Low Doses

Nearly all relevant cancer-risk data from human epidemiological studies and experimental animal bioassays reflect doses much higher than those typically experienced by humans from exposure to TCDD, other dioxins, and DLCs in the general environment. Consequently, analysts must extrapolate well below the doses observed in the studies to consider typical human exposure levels. This extrapolation involves two critical decisions: (1) selecting a "point of departure" (POD), which corresponds to the lowest dose associated with observable adverse effects within the range of data from a study, and (2) selecting the mathematical model used to extrapolate risk from typical human exposures that are well below the POD.

In general, EPA estimates the POD by setting it equal to the dose producing the smallest positive effect observed in a study. The size of the health effect it produces in the population determines the "effective dose." For example, the 1% effective dose (referred to as the ED_{01}) elicits an additional 1% response and the ED_{05} elicits an additional 5% response

[1]The charge to the committee was to evaluate EPA's Reassessment of dioxins and DLCs. Although other agencies, such as the International Agency for Research on Cancer (IARC), have also done both qualitative and quantitative evaluations of dioxin carcinogenicity, the committee focused solely on EPA's Reassessment document, the associated scientific evidence, and EPA's definitions for carcinogen classification.

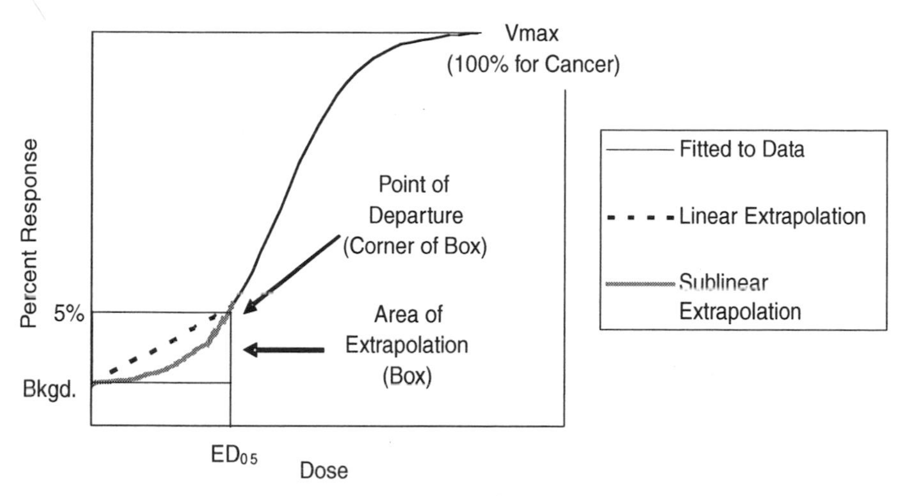

FIGURE S-1 Conceptual illustration of the effect of the selection of the point of departure and the mathematical model used to extrapolate below the point of departure on the risk estimate. Note that the 5% response rate is not drawn to scale. If it were, the area of the extrapolation box would be much smaller. In this illustration, the ED_{05} has been selected as the point of departure for extrapolation to lower doses.

above the "background" response (the level of response that occurs in the absence of any exposure). The response size depends on the difference between the unexposed population and the largest response possible. For example, consider the case of a 25% lifetime background risk of death from cancer in an unexposed population and a highest possible cancer death rate of 100%. In this case, the ED_{01} is the dose that increases the cancer death rate by 1% of the difference between 100% and 25%, or by 0.75%. Thus, the ED_{01} is the dose that increases the risk of dying from cancer from 25% to 25.75%.

Estimating risks below the POD requires making assumptions about how TCDD, other dioxins, and DLCs might cause cancer at lower exposures. For example, in the hypothetical illustration in Figure S-1, a biological mode of action implying that risk is proportional to dose would correspond to use of the dashed line below the POD. A biological mode of action implying a sublinear dose-response relationship would correspond to the shaded line below the POD.

The committee concludes that EPA's decision to rely solely on a default linear model lacked adequate scientific support. The report recommends that EPA provide risk estimates using both nonlinear and linear methods to extrapolate below PODs. If background exposures to humans result in doses substantially less than the dose associated with the POD (the most likely case in most instances but perhaps not for occupational exposures), then an estimate of risk for typical human exposures to TCDD, other dioxins, and DLCs would be lower in a sublinear extrapolation model than in the linear model. Given the important regulatory implications of this assumption, the committee recommends that EPA communicate the scientific strengths and weaknesses of both approaches so that the full range of uncertainty generated by modeling of the data is conveyed in the Reassessment.

The committee also concluded that EPA did not adequately quantify the uncertainty associated with responses at the estimated value of the POD. The estimated value of the response at a particular effective dose (like the ED_{01}) is typically uncertain for a variety of reasons related to the challenge of conducting an epidemiological study or an animal study. For example, in epidemiological studies, the number of enrolled subjects is limited, it can be difficult to estimate the actual level of exposure, other factors (such as smoking or exposure to other chemicals) can also cause cancer, and so forth. The committee concludes that, although EPA discussed many of these factors qualitatively, the agency should strive to more comprehensively characterize the impact of these sources of uncertainty quantitatively.

Estimating Noncancer Risk

To characterize the risks of adverse health effects other than cancer, EPA typically identifies a dose, called the reference dose (RfD), below which it anticipates no adverse effects from exposure even among sensitive members of the population. EPA did not estimate an RfD for TCDD, other dioxins, or DLCs in the Reassessment. The committee suggests that estimating an RfD would provide useful guidance to risk managers to help them (1) assess potential health risks in that portion of the population with intakes above the RfD, (2) assess risks to population subgroups, such as those with occupational exposures, and (3) estimate the contributions to risk from the major food sources and other environmental sources of TCDD, other dioxins, and DLCs for those individuals with high intakes.

Given the existing data, the committee concurs with the conclusion in EPA's Reassessment that TCDD, other dioxins, and DLCs are likely to be human immunotoxicants at "some dose level." However, the report finds this conclusion inadequate. The committee recommends that EPA add a section or paragraph to its Reassessment on the immunotoxicology of TCDD, other dioxins, and DLCs in the context of the biological mechanisms responsible for health effects relevant to assessing the likelihood of such effects occurring in humans at relatively low levels of exposure. The risk characterization should provide some insight about the level of risk given actual exposures.

Studies show that TCDD, other dioxins, and DLCs cause embryonic and fetal development and reproduction problems in rodents and some other species. However, the fetal rodent clearly shows more susceptibility to adverse effects of TCDD, other dioxins, and DLCs than the adult rodent. Given the lack of comparable human data, the committee recommends that EPA more thoroughly address how animal pregnancy models might relate to human reproductive and developmental toxicity and risk information.

The committee further recommends that, in areas with substantial amounts of human clinical data and epidemiological data, EPA establish formal, evidence-based approaches, including but not limited to those for assessing the quality of the study and study design for classifying and statistically reviewing all available data.

Communicating Variability and Uncertainty in Risk Estimates

Risk assessors must make many choices as they develop models to characterize risks, including selecting appropriate data sets for low-dose extrapolation, dose-response models, PODs, and so forth. Because risk estimates reflect numerous sources of uncertainty and alternative assumptions, EPA's Reassessment should include a detailed discussion of variability (the range of risks reflecting true differences among members of the population due to, for example, genetic and age differences) and uncertainty (the range of plausible risk estimates arising because of limitations in knowledge). Although EPA addressed many sources of variability and uncertainty qualitatively, the committee noted that the Reassessment would be substantially improved if its risk characterization included more quantitative approaches. Failure to characterize variability and uncertainty thoroughly can convey a false sense of precision in the conclusions of the risk assessment.

Estimating Toxicity of DLCs and Mixtures in the Environment

Risk managers base their decisions about cleanup and control of chemicals, such as TCDD, other dioxins, and DLCs, in the environment on assessment of the risks. Because of the common mode of action in producing health effects, EPA's Reassessment assessed the cumulative toxicity of the compounds. The approach taken by EPA and international public health organizations relies on assigning each compound (dioxins, other than TCDD, and DLCs) a "toxic equivalency factor," which is an estimate of the toxicity of the compound relative to TCDD. For example, a particular DLC thought to result in one-tenth the risk of TCDD for the same level of exposure would be assigned a toxicity equivalency factor of 0.1.

Because some mixtures may contain little or no measurable TCDD but relatively large amounts of other dioxins and/or DLCs, the toxic equivalency factor plays a critical role in determining the mixture's overall estimated toxicity (which is called the toxic equivalency quotient). Estimation of TEFs is a critically important part of the risk assessment of environmental mixtures of TCDD, other dioxins, and DLCs, because any environmental sample typically contains a dozen or more similar substances, but often very little TCDD. Also, TCDD, other dioxins, and DLCs break down at different rates in the environment and are eliminated at different rates in humans. Thus, although analysts may reasonably estimate the relative potency value for a given compound based on toxicity tests, the compound's contribution to total risk in an environmental (or biological) sample may change over a period of many years. This change may occur because the relative concentration in a sample may change with time, even though the potency remains constant, and the estimated risk in a given sample depends on both potency and concentration.

Even with the inherent uncertainties, the committee concludes that the toxic equivalency factor methodology provides a reasonable, scientifically justifiable, and widely accepted method to estimate the relative potency of DLCs. However, the committee noted that the Reassessment should acknowledge the need for better uncertainty analysis of the toxicity values and should provide at least some initial uncertainty analysis of overall toxicity of environmental samples.

CONCLUDING REMARKS

The committee appreciates the dedication and hard work that went into the creation of the Reassessment and commends EPA for its detailed evaluation of an extremely large volume of scientific literature (particularly Parts I and II of the Reassessment). The NRC report focused its review on Part III of the Reassessment and offers its recommendations with the intention of helping to guide EPA in its efforts to make and implement environmental policies that adequately protect human health and the environment from the potential adverse effects of TCDD, other dioxins, and DLCs. The committee recognizes that it will require a substantial amount of effort for EPA to incorporate all the changes recommended in this NRC report. Nevertheless, the committee encourages EPA to finalize the current Reassessment as quickly, efficiently, and concisely as possible after addressing the major recommendations in this report. The committee notes that new advances in the understanding of TCDD, other dioxins, and DLCs could require reevaluation of key assumptions in the EPA risk assessment document. The committee recommends that EPA routinely monitor new scientific information related to TCDD, other dioxins, and DLCs, with the understanding that future revisions should provide risk assessment based on the current state-of-the-science. However, the committee also recognizes the importance of stability in regulatory policy to the regulated community and thus suggests that EPA establish criteria for identifying when compelling new information warrants science-based revisions in its risk assessment. The committee finds that the recent dose-response data released by the National Toxicology Program after submission of the Reassessment represent good examples of new and compelling information that warrants consideration in a revised risk assessment.

COMMITTEE'S KEY FINDINGS

The committee identified three areas that require substantial improvement in describing the scientific basis for EPA's dioxin risk assessment to support a scientifically robust risk characterization:

- Justification of approaches to *dose response modeling* for cancer and noncancer end points.
- Transparency and clarity in *selection of key data sets* for analysis.
- Transparency, thoroughness, and clarity in *quantitative uncertainty analysis*.

The following points represent Summary recommendations to address the key concerns:

- EPA should compare cancer risks by using both a linear model and a nonlinear model consistent with a receptor-mediated mechanism of action and by using epidemiological data and the new NTP animal bioassay data. The comparison should include upper and lower bounds, as well as central estimates of risk. EPA should clearly communicate this information as part of its risk characterization.
- EPA should identify the most important data sets to be used for quantitative risk assessment for each of the four key end points (cancer, immunotoxicity, reproductive effects, and developmental effects). EPA should specify inclusion criteria for the studies (animal and human) used for derivation of the benchmark dose (BMD) for different noncancer effects and potentially for the development of RfD values and discuss the strengths and limitations of those key studies; describe and define (quantitatively to the extent possible) the variability and uncertainty for key assumptions used for each key end-point-specific risk assessment (choices of data set, POD, model, and dose metric); incorporate probabilistic models to the

extent possible to represent the range of plausible values; and assess goodness-of-fit of dose-response models for data sets and provide both upper and lower bounds on central estimates for all statistical estimates. When quantitation is not possible, EPA should clearly state it and explain what would be required to achieve quantitation.

- When selecting a BMD as a POD, EPA should provide justification for selecting a response level (e.g., at the 10%, 5%, or 1% level). The effects of this choice on the final risk assessment values should be illustrated by comparing point estimates and lower bounds derived from selected PODs.
- EPA should continue to use body burden as the preferred dose metric but should also consider physiologically based pharmacokinetic modeling as a means to adjust for differences in body fat composition and for other differences between rodents and humans.

The committee encourages EPA to calculate RfDs as part of its effort to develop appropriate margins of exposure for different end points and risk scenarios.

Summary

Dioxins are a class of chemicals, and the most toxic of these compounds is 2,3,7,8-tetrachlorodibenzo-*p*-dioxin (commonly referred to as TCDD or dioxin). There are many forms of dioxins and "dioxin-like compounds" (DLCs) that share most, if not all, of the toxic potential of TCDD, although nearly all are considerably less potent. Included in the list of DLCs are chlorinated forms of dibenzofurans and certain polychlorinated biphenyls (PCBs).

Combustion, metal processing, chemical manufacturing and processing, and other sources emit TCDD, other dioxins, and DLCs into the environment. Unlike PCBs, TCDD and other dioxins have never been intentionally produced. TCDD, other dioxins, and DLCs persist and bioaccumulate in the environment, which means that they break down slowly and build up through the food chain. Human exposure to TCDD, other dioxins, and DLCs occurs primarily from eating foods, such as beef, dairy products, fish, shellfish, and pork. In recent years, efforts to reduce the amount of TCDD, other dioxins, and DLCs in the environment have resulted in reductions in measured concentrations in the environment and in human blood.

TCDD, other dioxins, and DLCs share a common mode of action in producing toxic effects in humans and animals. They bind to a specific receptor, called the aromatic hydrocarbon receptor or Ah receptor; such binding is a necessary, but not sufficient, step toward producing adverse health effects.

A few industrial accidents and occupational exposures to substantial amounts of TCDD, other dioxins, and DLCs have provided opportunities to assess the toxicity of these compounds to humans. Several episodes of high-level human exposure to TCDD have been found to cause a specific type of persistent, potentially disfiguring skin lesion called chloracne. In 2004, the media widely publicized the suspected intentional poisoning of Viktor Yushchenko with TCDD after he developed chloracne during the Ukraine presidential campaign. In contrast to the undisputed high-dose effects of chloracne, the potential adverse effects of TCDD, other dioxins, and DLCs in humans after long-term, low-level environmental exposures remain controversial. The major controversies include how to classify the potential of these compounds to cause cancer in humans (as either "carcinogenic to humans" or "likely to be carcinogenic to humans"), how to estimate the potential health risks at very low doses typical of actual population exposures, and how to assess the toxicity of each of the compounds and various mixtures of them in the environment.

TCDD, other dioxins, and DLCs have been regulated extensively worldwide. In the early 1980s, the U.S. Environmental Protection Agency (EPA) and other organizations, such as the World Health Organization (WHO), began collecting and evaluating scientific information about the sources, fate, and effects of the compounds. In 1985, EPA produced an initial assessment of the human health risks from environmental exposure to TCDD. Later, as new scientific information became available, EPA reassessed the human health risks in an open process involving participation of numerous scientists external to the agency, a series of public meetings, and peer review.

An Interagency Working Group (IWG) made up of representatives of seven federal agencies was established in 2000 to coordinate federal strategies for risk management of TCDD, other dioxins, and DLCs. Members of the IWG, EPA's Science Advisory Board, and the public commented on earlier drafts of EPA's dioxin risk assessment, and after further revisions, EPA released the 2003 draft document titled *Exposure and Human Health Reassessment of Tetrachlorodibenzo-p-Dioxin (TCDD) and Related Compounds* (referred to as the Reassessment). The IWG recommended further review of the new document, and in 2004, EPA asked the National Research Council (NRC) to convene an expert committee to review independently EPA's 2003 draft Reassessment and to determine whether EPA's risk estimates are scientifically robust and whether there is clear delineation of all substantial uncertainties and variabilities (Box S-1).

This report presents the committee's conclusions and recommendations. In general, the committee recommends that EPA substantially augment its Reassessment to improve the transparency about assumptions used to estimate risk and how these assumptions affect estimates. The committee also recommends that EPA re-estimate the risks using several assumptions and communicate the uncertainty in these estimates to the public.

BOX S-1 Statement of Task

The National Academies' National Research Council will convene an expert committee that will review EPA's 2003 draft reassessment of the risks of dioxins and dioxin-like compounds to assess whether EPA's risk estimates are scientifically robust and whether there is a clear delineation of all substantial uncertainties and variability. To the extent possible, the review will focus on EPA's modeling assumptions, including those associated with the dose-response curve and points of departure; dose ranges and associated likelihood estimates for identified human health outcomes; EPA's quantitative uncertainty analysis; EPA's selection of studies as a basis for its assessments; and gaps in scientific knowledge. The study will also address the following aspects of the EPA reassessment: (1) the scientific evidence for classifying dioxin as a human carcinogen; and (2) the validity of the non-threshold linear dose-response model and the cancer slope factor calculated by EPA through the use of this model. The committee will also provide scientific judgment regarding the usefulness of toxicity equivalence factors (TEFs) in the risk assessment of complex mixtures of dioxins and the uncertainties associated with the use of TEFs. The committee will also review the uncertainty associated with the reassessment's approach regarding the analysis of food sampling and human dietary intake data and, therefore, human exposures, taking into consideration the Institute of Medicine's report *Dioxin and Dioxin-Like Compounds in the Food Supply: Strategies to Decrease Exposure*. The committee will focus particularly on the risk characterization section of EPA's reassessment report and will endeavor to make the uncertainties in such risk assessments more fully understood by decision makers. The committee will review the breadth of the uncertainty and variability associated with risk assessment decisions and numerical choices—for example, modeling assumptions, including those associated with the dose-response curve and points of departure. The committee will also review quantitative uncertainty analyses, as feasible and appropriate. The committee will identify gaps in scientific knowledge that are critical to understanding dioxin reassessment.

CARCINOGENIC CLASSIFICATION

In 1985, EPA classified TCDD as a "probable human carcinogen" based on the data available at the time, but the latest Reassessment (2003) stated that TCDD was better characterized as "carcinogenic to humans." EPA and the International Agency for Research on Cancer (IARC), an arm of WHO, have established criteria for qualitatively classifying chemicals into various carcinogenic categories based on the weight of scientific evidence from animal, human epidemiological, and mechanism or mode-of-action studies. In 1997, an expert panel convened by IARC concluded that the weight of scientific evidence for dioxin carcinogenicity in humans supported its classification as a Class 1 carcinogen—"carcinogenic to humans."

In 2001, the U.S. National Toxicology Program (NTP) upgraded its classification of dioxin to "known to be a human carcinogen."

After reviewing EPA's 2003 Reassessment and other scientific information and in light of EPA's recently revised 2005 *Guidelines for Carcinogen Risk Assessment* (cancer guidelines), the committee concludes that the classification of TCDD as "carcinogenic to humans"—a designation suggesting the greatest degree of certainty about carcinogenicity—versus "likely to be carcinogenic to humans"—the next highest designation—is somewhat subjective and depends largely on the definition and interpretation of the criteria used for classification. The true weight of evidence lies on a continuum, with no obvious point or "bright line" that readily distinguishes those two categories.

Referring to the specific definitions in EPA's 2005 cancer guidelines for qualitative classification of chemical carcinogens, the NRC committee was split on whether the evidence met *all* the criteria necessary for classification of TCDD as "carcinogenic to humans," although the committee unanimously agreed on a classification of at least "likely to be carcinogenic to humans." The committee concludes that the weight of epidemiological evidence supporting classification of TCDD as a human carcinogen is not "strong." The committee points out, however, that the human data available from occupational studies show a modest positive association between relatively high concentrations of TCDD in the body and increased mortality from all cancers. Animal studies and mechanistic data provide additional support for classifying TCDD as a human carcinogen.

The committee concludes that the distinction between those two qualitative categories of cancer risk classification depends more on semantics than on science and that the *public health* implications of the two terms appeared identical, and for these reasons the committee did not focus much attention on the issue of classification. To the extent that EPA can be consistent with regulatory requirements, the committee recommends that EPA focus its energies and resources on more carefully quantifying risks and uncertainties for TCDD, other dioxins, and DLCs rather than on whether its carcinogenicity is probable or proven. Because the 2005 cancer guidelines' definition of "carcinogenic to humans" has changed since EPA completed its 2003 Reassessment, the committee recommends that EPA reevaluate its conclusion that TCDD satisfies the criteria for designation as either "carcinogenic to humans" or "likely to be a human carcinogen" based on the criteria set out in EPA's 2005 cancer guidelines.

The committee agrees with EPA in classifying dioxins, other than TCDD, and DLCs as "likely to be carcinogenic to humans." However, because mixtures of DLCs may also contain dioxins, including TCDD, EPA should reconsider its classification of such mixtures as "likely to be carcinogenic to humans" if it continues to classify dioxin as "carcinogenic to humans."

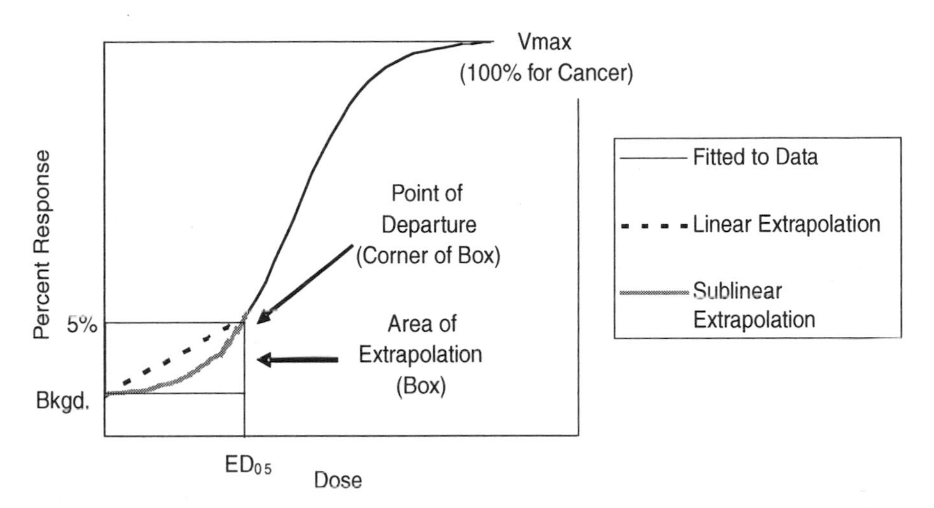

FIGURE S-1 Conceptual illustration of the effect of the selection of the point of departure and of the mathematical model used to extrapolate below the point of departure on the risk estimate. Note that the 5% response rate is not drawn to scale. If it were, the area of extrapolation box would be much smaller.

ESTIMATING CANCER RISK

Because nearly all data (both human epidemiological studies and experimental animal bioassays) relevant to cancer risk are for doses much higher than those to which the general human population is typically exposed, analysts must extrapolate below the doses observed when estimating risks. This extrapolation depends on first fitting a dose-response curve to the observed data from a given study and choosing a "point-of-departure" (POD) dose, which corresponds to the lowest dose associated with adverse effects within the range of the data from the experiment or study. The POD dose is an incremental "effect" observed; for example, analysts would call a POD corresponding to a 5% increase in effects (above no exposure) a 5% effective dose or an ED_{05}.

Estimating risks below the POD may require extrapolating down to background levels of exposure. See Figure S-1 for a conceptual illustration of a dose-response extrapolation to background levels using the 5% response rate and ED_{05} as the POD. This extrapolation must be based on assumptions about how TCDD, other dioxins, and DLCs might cause cancer. Thus, the selection of the type of mathematical model used to extrapolate below the POD is a critical decision in the cancer risk assessment process. In the 2003 Reassessment, EPA chose to extrapolate below the POD with a "linear" model, which assumes that the biological response increases proportionally with the level of exposure starting at a dose of zero. Risk estimates based on this approach are generally higher than those based on alternative "nonlinear" assumptions, where the biological response does not vary proportionally with the dose. However, EPA took the position that scientific data were inadequate to rule out its default linear assumption.

Selection of the POD also is an important choice in cancer risk assessment modeling because it determines the range of extrapolation below the observed data range. For example, more extrapolation below the POD is necessary when using a POD equal to the ED_{10} than when using a POD equal to the ED_{01}. However, using an ED_{01} requires more data because the analyst must be able to detect a 1% increase in effects instead of a larger increase (e.g., a 5% increase for an Ed_{05}).

After reviewing EPA's 2003 Reassessment and additional scientific data published since completion of the Reassessment, the committee unanimously agreed that the current weight of scientific evidence on the carcinogenicity of dioxin is adequate to justify the use of nonlinear methods consistent with a receptor-mediated response to extrapolate below the POD. The committee points out that data from NTP released after EPA generated the 2003 Reassessment provide the most extensive information collected to date about TCDD carcinogenicity in test animals, and the committee found the NTP results to be compelling. The committee concludes that EPA should reevaluate how it models the dose-response relationships for TCDD, other dioxins, and DLCs. Specifically, the committee determined that the scientific evidence is consistent with receptor-mediated responses and favors the use of a nonlinear model over the default linear assumption to extrapolate below the POD for dioxin-related cancer risk. The committee recognizes that a linear response at doses below the POD cannot be entirely excluded, especially if background exposures are not orders of magnitude below the POD and additivity of risk from other types of chemicals is considered.

Because the committee concludes that the data support the hypothesis that the dose-response relationship for dioxin and cancer is sublinear, it recommends that EPA include a nonlinear model for cancer risk estimates but also use the current linear models for comparative purposes. EPA should then describe the scientific strengths and weaknesses of each approach to inform risk managers about the importance of these assumptions. The committee recognizes that additional evidence about dioxin carcinogenicity will continue to develop and concludes that EPA should proceed with completing its quantitative cancer risk assessment and include the recent NTP data and appropriate nonlinear dose-response models.

ESTIMATING NONCANCER RISK

To characterize the risks of adverse health effects other than cancer at very low doses, EPA typically identifies a dose called the reference dose (RfD) below which it anticipates no adverse effects from exposure, even among sensitive members of the population. To estimate the RfD, EPA usually starts with a benchmark dose (BMD), which is the level of exposure in an epidemiological study or an animal experiment that produces a certain specified level of response. For example, the BMD might be defined as the level of exposure at which 5% of exposed animals or people exhibit a specific type of adverse effect. EPA then calculates the RfD by dividing the BMD by a series of uncertainty factors intended to take into account several sources of uncertainty. These sources of uncertainty include extrapolation from animals to humans (allowing for a more sensitive response in humans than that in the test animals), extrapolation within the population (allowing for more sensitive members of the human population), and database sufficiency considerations (allowing for the possibility that more data might reveal more sensitive effects).

EPA did not estimate an RfD for TCDD, other dioxins, and DLCs in the Reassessment. However, the committee noted that defining an RfD would provide useful guidance to risk managers to help them (1) assess potential health risks in that portion of the population with intakes above the RfD, (2) assess risks to population groups, such as those with occupational exposures, and (3) estimate the risk contributions of the major food sources and other environmental sources for those individuals with high intakes. Alternatively, EPA could undertake risk characterization for different adverse effects by comparing noncancer dose-response data to relevant human exposure data in the calculation of margins of exposure (MOEs),[1] as was done in the Reassessment. Such MOEs, accompanied by a description of associated uncertainties, could provide risk managers with information that would help to inform their decisions. Although EPA concluded that calculating RfDs would not provide useful information, the committee concludes that the information might be useful if EPA also considered that the use of body burden (estimate of the total amount of chemical in the body at steady state for a defined rate of exposure) as a dose metric would already take into account some of the uncertainty factors that EPA would typically use to adjust the BMD or POD in estimating an RfD. Estimates of background exposures in the United States also appear to have continued to decline, in part due to enhanced analytical detection.

[1]EPA defines MOE as the lowest "ED_{10} or other point of departure divided by the actual or projected environmental exposure of interest" (http://www.epa.gov/iris/gloss8.htm#m).

The committee concludes that EPA did not adequately justify the use of the 1% response level (the ED_{01}) as the POD for analyzing epidemiological or animal bioassay data for both cancer and noncancer effects. The committee recommends that EPA more explicitly address the importance of the selection of the POD and its impact on risk estimates by calculating risk estimates using alternative assumptions (e.g., the ED_{05}).

The committee commends EPA's extensive dose-response modeling efforts of a large number of data sets, particularly those for noncancer effects, but remains concerned about selection of the final model for computing the POD. It is critical that the model used for determining a POD fit the data well, especially at the lower end of the observed responses. Whenever feasible, mechanistic and statistical information should be used to estimate the shape of the dose-response curve at lower doses. At a minimum, EPA should use rigorous statistical methods to assess model fit, and to control and reduce the uncertainty of the POD caused by a poorly fitted model. The overall quality of the study design is also a critical element in deciding which data sets to use for quantitative modeling.

UNCERTAINTY AND VARIABILITY IN RISK ESTIMATES

Risk assessors must make many choices as they develop models to characterize risks. Some of the initial choices are selecting appropriate data sets for low-dose extrapolation, selecting appropriate dose-response models, selecting critical end points, and selecting an appropriate POD (e.g., ED_{01} versus ED_{05}). Risk estimates routinely reflect numerous sources of both uncertainty (which describes the range of plausible risk estimates arising because of limitations in knowledge) and variability (which describes the range of risks arising because of true differences—for example, genetic and age differences among members of the population). Failure to fully characterize uncertainty and variability can convey a false sense of precision in the conclusions of the risk assessment. EPA should include a detailed discussion of both uncertainty and variability.

Overall, the committee concludes that EPA addressed many sources of uncertainty and variability qualitatively, but it did not adequately quantify either the uncertainty or the variability of many. In the case of its cancer risk estimates, EPA should provide quantitative estimates corresponding to (1) central, upper-bound, and lower-bound estimates of the POD; (2) the use of different plausible POD values; (3) different plausible mathematical functions fit to the observed epidemiological data; and (4) different assumptions for estimating historical exposures among subjects in the epidemiological studies. In the case of the noncancer risk estimates, EPA should characterize the uncertainty associated with (1) fitting a dose-response relationship to the available data and (2) selecting a POD. If necessary, EPA should acknowledge that the information available remains insufficient to support a meaningful point estimate.

EPA's discussion of epidemiological studies in Part III of the Reassessment, *Integrated Summary and Risk Characterization for TCDD and Related Compounds*, should clearly specify inclusion criteria for those studies used as a basis to support quantitative risk estimates. The committee notes that EPA could substantially improve the transparency and management of the uncertainties and complexities of the risk assessment for TCDD, other dioxins, and DLCs by creating an ongoing process for clearly identifying and updating the key assumptions that support the quantitative risk assessment.

ESTIMATING TOXICITY OF DLCS AND MIXTURES

Many DLCs and dioxins, other than TCDD, present in the environment are capable of producing toxicological effects similar or identical to those of TCDD. Substantial efforts have been aimed at simplifying estimation of risk for these compounds and for mixtures of them. EPA and international public health organizations have tended to take the approach of assigning each compound (dioxins, other than TCDD, and DLCs) a toxic equivalency factor (TEF), which represents a scaling factor for estimating the toxicity of the compound relative to TCDD. For example, a substance with a TEF of 0.1 is estimated to be 10% as toxic as dioxin per unit mass. Estimation of TEFs is a critically important part of the risk assessment of environmental mixtures of TCDD, other dioxins, and DLCs, because any environmental sample typically contains a dozen or more similar substances, but often very little TCDD. TCDD, other dioxins, and DLCs break down at different rates in the environment and have different elimination rates in humans. Thus, although analysts may reasonably estimate the relative potency value for a given compound based on toxicity tests, the compound's contribution to total risk in an environmental (or biological) sample may change over a period of many years. This change may occur because the relative concentration in a sample may change with time, even though the potency remains constant, and the estimated risk in a given sample depends on both potency and concentration. Because these mixtures may contain little or no TCDD but relatively large amounts of low-potency dioxins and/or DLCs, TEFs are a critical factor in determining the mixture's overall estimated toxicity. Analysts refer to the aggregate weighting by TEF of a mixture as the mixture's toxic equivalent quotient (TEQ).

The recent NTP studies on TCDD and several other dioxins and DLCs provide additional evidence in support of the TEF approach. Uncertainty about the validity of the approach led the NTP to specifically test the TEF value for one particular PCB (126) in its analyses, and the results showed excellent agreement between the predicted TEF for PCB 126 and the value observed in the NTP experiment.

Overall, even given the inherent uncertainties, the committee agrees that the TEF method is reasonable, scientifically justifiable, and widely accepted for the estimation of the relative toxic potency of TCDD, other dioxins, and DLCs. The TEF approach has also been used in other contexts. WHO's International Programme on Chemical Safety used the approach in assessing the risks of different polycyclic aromatic hydrocarbons (PAHs) relative to benzo(*a*)pyrene as an indicator PAH.

The committee concludes that the recent NTP results, released after EPA completed its 2003 Reassessment, provide important additional support for the TEF approach. However, EPA should acknowledge the need for better uncertainty analysis of the TEF values and should, as a follow-up to the Reassessment, establish a task force to begin to address this uncertainty by developing "consensus probability density functions" for dioxins and DLCs. The committee recommends that EPA clearly address TEF uncertainties in the Reassessment.

SCALING DATA FROM ANIMAL STUDIES

For risk assessments that rely on experimental animal data, determining the most appropriate way to scale the data from the animal model (usually rats and mice) to humans is another important risk assessment choice. Numerous options for choosing dose metrics exist, and they can yield results different from the traditional daily dose metric based on per unit of body weight. For highly persistent chemicals like TCDD, other dioxins, and DLCs, substantial differences in the rates of elimination from the body will result in very different amounts of chemical accumulated in the body over time, even with the same daily dose rate expressed in body weight or body surface area units. In the 2003 Reassessment, EPA used an estimate of the total amount of chemical in the body at steady state for a defined rate of exposure, called the body burden, as the dose metric to adjust for differences in body weight (or surface area) and in elimination rates.

The committee agrees with EPA's conclusion that use of body burden as the dose metric appears to be the most reasonable and pragmatic approach for dioxin risk assessment, but EPA should address important uncertainties quantitatively in more detail when possible. One such uncertainty, not quantitatively addressed in EPA's 2003 Reassessment, relates to species differences in body fat expressed as a percentage of total body weight. Differences in body fat content have a potentially large impact on dioxin concentrations present in nonfatty tissues, including such organs as the liver.

Large errors may also arise from trying to estimate the overall body burden TEQs for humans based on intake TEFs from rats. The errors result from uncertainties in the differences in how long TCDD, other dioxins, and DLCs persist in humans and rodents and uncertainties about how these compounds concentrate in tissues.

The committee recommends that EPA's Reassessment use basic physiologically based pharmacokinetic (PBPK) models to estimate the differences between humans and rodents in the relationship between total body burden at steady state, as calculated from the intake, half-life, bioavailability, and tissue concentrations, and use the results to modify the estimated human equivalent intakes. The committee also recommends that EPA provide a clear evaluation of the impact of using body burden as the dose metric, relative to other possible options such as intake, on the final risk estimates.

HUMAN EXPOSURE TO TCDD, OTHER DIOXINS, AND DLCS

Estimating human exposure levels, including those representative of background levels (e.g., typical dietary intake levels) and levels resulting from specific exposure scenarios (e.g., accidental, occupational, and highly exposed communities), is a critical component of any chemical risk assessment. The extensive environmental persistence of TCDD, other dioxins, and DLCs and their global environmental distribution create many possible sources and routes of exposure to these compounds, and determining typical background rates of exposure is difficult. EPA's 2003 Reassessment addresses exposure to TCDD, other dioxins, and DLCs in terms of sources, environmental fate, environmental media concentrations, food concentrations, background exposures, and potentially highly exposed populations.

To assess total dioxin and DLC emissions, EPA used a "bottom-up" approach in which it attempted to identify all source categories and then estimated the emissions for each category. However, a "top-down" approach that attempts to account for measured levels and considers the emission sources required to account for those levels would provide useful additional information. Such alternative approaches may give rise to significantly different estimates of the historical levels of dioxin and DLC emissions. Both approaches come with uncertainties, and EPA could benefit substantially from using the approaches simultaneously to set plausible bounds on historical trends and current levels in emissions.

The committee also recommends that EPA more explicitly define its procedures for addressing analytical measurements that fall below the limit of detection in environmental and exposure media samples. Consideration of the detection limits is important in assessing background exposure estimates. Typically, samples that contain small or no amounts of dioxin ("nondetects") are given a value of 50% of the lowest level measurable by the instrument (the detection limit). For example, if the detection limit was 1 part per billion (ppb), a sample that contained 0.1 ppb would be assigned a value of 0.5 ppb (half of the detection limit), or 5 times greater than its actual value. If the detection limit decreased to 0.05 ppb, the actual value of 0.1 ppb would be reported. In addition, as analytical detection limits improve, the estimates of contaminants in background environmental samples become more accurate as nondetect samples become fewer and the range of uncertainty is narrowed.

Although beyond the scope of the review of the EPA Reassessment, the committee notes that it would be useful for EPA to set up a compound-specific, active database of *typical* concentrations for the range of TCDD, other dioxins, and DLCs present in dietary and other environmental sources. This database should undergo regular updates to capture new data as they appear in the peer-reviewed literature. Such a database should include clear requirements of data quality and traceability (chemical analysis, representative and targeted sampling, representative of consumer exposure, presentation of data, and handling and presentation of "nondetects"). Chapter 4 provides several additional recommendations about the exposure assessment section of the 2003 Reassessment.

IMMUNOTOXICITY OF TCDD, OTHER DIOXINS, AND DLCS

TCDD, other dioxins, and DLCs have well-known effects on the immune systems of experimental animals. Chemically induced alterations in immune function could result in various adverse health outcomes because the immune system plays a critical role in fighting off infections, killing cancer cells at early stages, and implementing numerous other health-protective functions.

In light of the large database showing that TCDD, other dioxins, and DLCs produce immunotoxic responses in laboratory animal studies, combined with sparse human data, the committee agrees with EPA's conclusion that these compounds are potential human immunotoxicants.

However, EPA's conclusion that dioxins, other than TCDD, and DLCs are immunotoxic at "some dose level" is inadequate. At a minimum, EPA should add a section or paragraph that discusses the immunotoxicology of these compounds in the context of current Ah receptor biology. EPA should also include some discussion about the implications of using genetically homogeneous inbred mice to characterize immunotoxicological risk in the genetically variable human population.

REPRODUCTIVE AND DEVELOPMENTAL TOXICITY OF TCDD, OTHER DIOXINS, AND DLCS

Reproduction and embryonic and fetal development are sensitive end points from rodent exposure to TCDD, other dioxins, and DLCs. Although the fetal rodent consistently appears to be more susceptible to adverse effects of these compounds than the adult rodent, comparable human data do not exist, and the susceptibility of humans to these end points is less well determined.

EPA's 2003 Reassessment comprehensively covers developmental and reproductive toxicity of TCDD, other dioxins, and DLCs in several models. One rodent model included TCDD administration during pregnancy and thus tested the disruption of development of the pups and their reproductive function later in life. The 2003 Reassessment presented a comprehensive overview of the pregnancy model, but it did not provide an adequate discussion of the doses used in the studies or the relationships of animal studies to human reproductive and developmental toxicity. The committee recommends that EPA more thoroughly address how the effective doses used in the animal pregnancy models relate to human reproductive and developmental toxicity and risk information, including TEFs and TEQs. The 2003 Reassessment also did not provide an adequate discussion of other models (e.g., effects of TCDD on ovulation in adult rats).

OTHER TOXIC END POINTS

Although TCDD, other dioxins, and DLCs have received wide recognition for their potential to cause cancer, birth defects, reproductive disorders, immunotoxicity, and chloracne, animal and human studies have demonstrated other potential toxic end points, including liver disease, thyroid dysfunction, lipid disorders, neurotoxicity, cardiovascular disease, and metabolic disorders, such as diabetes.

The committee agrees that EPA has in general adequately addressed the available data on the likelihood that exposure to TCDD, other dioxins, and DLCs is a significant risk factor for other toxic end points. EPA cautiously stated its overall conclusions about noncancer risks due to TCDD, other dioxins, and DLCs exposures and acknowledged the uncertainty of suspected relationships. Nonetheless, the committee notes that EPA did not uniformly address the limitations of individual human studies. Similarly, EPA did not discuss the broad 95% confidence intervals accompanying some reported statistically significant effects in the context of the uncertainty (and, perhaps, individual variability) that these broad confidence limits imply. Conversely, the 2003 Reassessment highlights statistically nonsignificant effects in some cases, suggesting an implied potential for unobserved detrimental effects without a supporting presentation of a firm evidence base. The committee recommends that EPA establish formal principles and mechanisms for evidence-based classification and systematic statistical review, including meta-analysis when possible, for available human, clinical, and noncancer end-point data.

New studies of the effects of dioxin on the developing vascular system suggest a potentially sensitive target for TCDD, other dioxins, and DLCs. The committee recommends that EPA identify this area as an important data gap in the understanding of the potential adverse effects of these compounds.

EPA'S OVERALL APPROACH TO RISK CHARACTERIZATION

Risk characterization is the culminating step in risk assessment. It should attempt to pull together all the relevant scientific information on toxicity and exposure for a coherent, quantitative understanding of potential health risks and on the uncertainties that surround the estimates of risk. Ideally, the risk characterization component of a risk assessment provides risk managers with a user-friendly synopsis of the scientific basis that underpins an agent's potential impact on public health under defined exposure conditions and scenarios.

As discussed previously, selection of the default linear extrapolation approach for carcinogenicity emerged as one of the most critical decisions in the 2003 Reassessment. The committee concludes that EPA did not support its decision adequately to rely solely on this default linear model and recommends that EPA add a scientifically rigorous evaluation of a nonlinear model that is consistent with receptor-mediated responses and uses the recent NTP cancer bioassay studies. The committee determined that the available data support the use of a nonlinear model, which is consistent with receptor-mediated responses and a potential threshold, with subsequent calculations and interpretation of MOEs. EPA's sole use of the default assumption of linearity and selection of the ED_{01} as the only POD to quantify cancer risk does not provide an adequate quantitative characterization of the overall range of uncertainty associated with the final estimates of cancer risk.

Because EPA decided not to derive an RfD, its traditional noncancer metric, or any other alternative for noncancer effects, the 2003 Reassessment does not provide important detailed risk characterization information about noncancer risks. Typically, when EPA estimates an RfD, the risk characterization will include (1) estimates of the proportion of the population with intakes above the RfD; (2) detailed assessment of population groups, such as those with occupational exposures; and (3) contributions of the major food sources and other environmental sources for those individuals with high intakes. If a nonlinear model consistent with a threshold were used for cancer risk assessment, these same types of risk characterization details could also be provided for cancer risk. The lack of such a focus in the risk characterization section of the 2003 Reassessment results in a risk characterization that is difficult to follow and does not provide clear guidance with respect to noncancer end points.

The committee recommends that EPA revise its risk characterization chapter to clearly describe the following:

1. The effects seen at the lowest body burdens that are the primary focus for any risk assessment—the "critical effects."

2. The modeling strategy used for each noncancer effect modeled, paying particular attention to the critical effects, and the selection of a point of comparison based on the biological significance of the effect; if the ED_{01} is retained, then the biological significance of the response should be defined and the precision of the estimate given.

3. The precision and uncertainties associated with the body burden estimates for the critical effects at the point of comparison, including the use of total body burden rather than modeling steady-state concentrations for the relevant tissue.

4. The committee encourages EPA to calculate RfDs as part of its effort to develop appropriate margins of exposure for different end points and risk scenarios, including the proportions of the general population and of any identified groups that might be at increased risk (See Table A-1 in the Reassessment, Part III Appendix, for the different effects; appropriate exposure information would need to be generated.) Interpretation of the calculated values should take into consideration the uncertainties in the POD values and intake estimates.

5. Consideration of individuals in susceptible life stages or groups (e.g., children, women of childbearing age, and nursing infants) who might require estimation of a separate MOE using specific exposure data.

6. Distributions that provide clear insights about the uncertainty in the risk assessments, along with discussion of the key contributors to the uncertainty.

The committee recommends that EPA substantially revise the risk characterization section of Part III of the Reassessment to include a more comprehensive risk characterization and discussion of the uncertainties surrounding key assumptions and variables.

CONCLUDING REMARKS

The committee appreciates the dedication and hard work that went into the creation of the Reassessment and commends EPA for its detailed evaluation of an extremely large volume of scientific literature (particularly Parts I and II of the Reassessment). This NRC report focuses its review on

Part III of the Reassessment and offers its recommendations with the intention of helping to guide EPA in its efforts to make and implement environmental policies that adequately protect human health and the environment from the potential adverse effects of TCDD, other dioxins, and DLCs. The committee recognizes that it will require a substantial amount of effort for EPA to incorporate all the changes recommended in this report. Nevertheless, the committee encourages EPA to finalize the current Reassessment as quickly, efficiently, and concisely as possible after addressing the major recommendations in this report. The committee notes that new advances in the understanding of TCDD, other dioxins, and DLCs could require reevaluation of key assumptions in EPA's risk assessment document. The committee recommends that EPA routinely monitor new scientific information related to TCDD, other dioxins, and DLCs, with the understanding that future revisions may be required to maintain a risk assessment based on the current state-of-the-art science. However, the committee also recognizes that stability in regulatory policy is important to the regulated community and therefore suggests that EPA establish criteria for identifying when compelling new information would warrant science-based revisions in its risk assessment. The committee finds that the recent dose-response data released by the NTP after submission of the Reassessment are good examples of new and compelling information that warrants consideration in a revised risk assessment.

COMMITTEE'S KEY FINDINGS

The committee identified three areas that require substantial improvement in describing the scientific basis for EPA's dioxin risk assessment to support a scientifically robust risk characterization:

- Justification of approaches to *dose-response modeling* for cancer and noncancer end points.
- Transparency and clarity in *selection of key data sets* for analysis.
- Transparency, thoroughness, and clarity in *quantitative uncertainty analysis.*

The following points represent Summary recommendations to address the key concerns:

- EPA should compare cancer risks by using nonlinear models consistent with a receptor-mediated mechanism of action and by using epidemiological data and the new NTP animal bioassay data. The comparison should include upper and lower bounds, as well as central estimates of risk. EPA should clearly communicate this information as part of its risk characterization.

- EPA should identify the most important data sets to be used for quantitative risk assessment for each of the four key end points (cancer, immunotoxicity, reproductive effects, and developmental effects). EPA should specify inclusion criteria for the studies (animal and human) used for derivation of the benchmark dose (BMD) for different noncancer effects and potentially for the development of RfD values and discuss the strengths and limitations of those key studies; describe and define (quantitatively to the extent possible) the variability and uncertainty for key assumptions used for each key end-point-specific risk assessment (choices of data set, POD, model, and dose metric); incorporate probabilistic models to the extent possible to represent the range of plausible values; and assess goodness-of-fit of dose-response models for data sets and provide both upper and lower bounds on central estimates for all statistical estimates. When quantitation is not possible, EPA should clearly state it and explain what would be required to achieve quantitation.
- When selecting a BMD as a POD, EPA should provide justification for selecting a response level (e.g., at the 10%, 5%, or 1% level). The effects of this choice on the final risk assessment values should be illustrated by comparing point estimates and lower bounds derived from selected PODs.
- EPA should continue to use body burden as the preferred dose metric but should also consider physiologically based pharmacokinetic modeling as a means to adjust for differences in body fat composition and for other differences between rodents and humans.

1

Introduction

The U.S. Environmental Protection Agency (EPA) and other organizations, such as the World Health Organization (WHO), began assessing the potential risks to human health from exposure to 2,3,7,8-tetrachlorodibenzo-*p*-dioxin (TCDD, commonly referred to as dioxin) decades ago. Early studies suggested very high toxicity at very low doses in test animals and potential carcinogenicity. The history of dioxin risk assessment is complicated and contentious (Thompson and Graham 1997). In 1985, EPA produced an initial assessment of the human health risks from environmental exposure to dioxin. Three years later, EPA, other federal agencies, and the scientific community began developing a broad research program to identify the biological response mechanisms and to explore other key scientific issues related to dioxin. In light of significant advances in the scientific understanding of the mechanisms of dioxin toxicity, new studies of dioxin's carcinogenic potential in humans, and increased evidence of other adverse health effects primarily after the 1985 assessment, EPA announced in 1991 that it would conduct a scientific reassessment of the health risks of human exposure to TCDD and related compounds, that is, dioxins, other than TCDD, and dioxin-like compounds (DLCs). The reassessment would respond to emerging scientific knowledge of the biological, human health, and environmental effects of TCDD, other dioxins, and DLCs.

EPA conducted the reassessment process as an open and participatory exercise, involving chapter authorship by scientists outside the agency, a series of public meetings and peer-review workshops, and reviews by EPA's Science Advisory Board (SAB). EPA's National Center for Environmental

Assessment (NCEA) headed the reassessment efforts with participation of scientific experts in EPA, the National Institutes of Health's (NIH) National Institute of Environmental Health Sciences (NIEHS), and other federal agencies and scientific experts in the private sector and academia. EPA sponsored open meetings in 1991 and 1992 to inform the public about the assessment, receive public comments on plans and activities of the reassessment process, and obtain additional relevant scientific information. Peer-review workshops were convened in 1992 and 1993 to review initial drafts of all background chapters. The workshops were followed by extensive revision and additional review of some chapters. In 1994, EPA released for public review all the chapters plus the first draft of a summary risk characterization chapter, received public comments on the drafts, and submitted the documents to the SAB for review.

In 1995, the SAB, commenting on the 1994 draft assessment, proposed several substantive and contingent recommendations, including revision of the chapter on dose-response modeling for TCDD, development of a chapter on dioxin toxic equivalency factors (TEFs), and an external peer review of redrafted or new chapters, including the chapter on risk characterization. The SAB also recommended that EPA involve outside scientists from the public and private sectors to help determine approaches for revising what was then called Chapter 9: "Risk Characterization of TCDD and Related Compounds."

In 1996, EPA initiated interaction with a group of 40 stakeholders from the public and private sectors to gather input on approaches for conducting the risk characterization revision. EPA met regularly with the group to ensure ongoing input as recommended by the SAB and shared with them the initial post-SAB revision of the draft risk characterization.

EPA, with NIEHS, revised Chapter 8, developed a new Chapter 9 on TEFs, and revised the former Chapter 9 and renamed it as a free-standing report, "Part III—Integrated Summary and Risk Characterization for 2,3,7,8-Tetrachlorodibenzo-*p*-Dioxin (TCDD) and Related Compounds." The dioxin report consisted of two other parts: "Part I—Estimating Exposure to Dioxin-Like Compounds" and "Part II—Health Assessment for 2,3,7,8-Tetrachlorodibenzo-*p*-Dioxin (TCDD) and Related Compounds." All three parts are collectively referred to as the Reassessment.

On February 24, 1997, the *Federal Register* announced the public external peer review and 60-day comment period of the revised Chapter 8, "Dose-Response Modeling for 2,3,7,8-TCDD." On June 12, 2000, the *Federal Register* announced a similar peer review and public comment period on the revised Part III—Integrated Summary and Risk Characterization and the revised TEFs Chapter 9 in Part II.

Recognizing the broad policy implications of the dioxin reassessment, the National Science and Technology Council established an interagency working group on dioxin (IWG) in the summer of 2000 to ensure a coordinated federal approach to dioxin-related health, food, and environmental issues. Specifically, the IWG was charged with fostering information sharing, developing a common language for dioxin science and science policy across governmental agencies and programs, identifying gaps and needs in the dioxin risk assessment, and facilitating coordination of risk management strategies. The IWG includes representatives from the following federal agencies: U.S. Department of Health and Human Services, U.S. Department of Agriculture, U.S. Department of State, U.S. Department of Veterans Affairs, U.S. Department of Defense, the Executive Office of the President, and EPA.

In the winter of 2000, the SAB held a 3-day public review of Part III of the Reassessment and additional information on the toxic equivalence of dioxins, other than TCDD, and DLCs. In the spring of 2001, the SAB recommended that EPA proceed expeditiously to complete and release its report, taking appropriate note of the SAB's findings and recommendations and public comments. In response, EPA revised its draft Reassessment and submitted it to the IWG in late 2003, requesting input about the need and benefit of further review. EPA appropriations language for fiscal year 2003 also called for an IWG evaluation of the need for further review and provided specific issues to consider. The IWG recommended that the National Academies' National Research Council (NRC) review the draft Reassessment. The scope of work for the NRC review and interagency agreements for funding were developed through the IWG in the spring of 2004. Ultimately, the NRC review would seek to inform and assure the risk characterization of TCDD, other dioxins, and DLCs and to benefit EPA in finalizing its Reassessment.

TCDD, OTHER DIOXINS, AND DLCS

The Reassessment addresses a limited number of chemical compounds within three subclasses of the halogenated aromatic hydrocarbons (HAHs): the polychlorinated dibenzo-*p*-dioxins (PCDDs), the polychlorinated dibenzofurans (PCDFs), and the polychlorinated biphenyls (PCBs). These compounds contain the basic aromatic structure of a benzene ring, a hexagonal carbon structure with conjugated double bonds connecting the carbons (Figure 1-1). PCDDs and PCDFs have tricyclic (triple-ring) structures consisting of two benzene rings, with varying numbers of chlorines, connected by an oxygenated ring, with the oxygenated ring of PCDDs having two oxygen atoms (a dioxin, Figure 1-2a) and the oxygenated ring of PCDFs having a single oxygen atom (a furan, Figure 1-2b). PCBs have a variable number of chlorines attached to a biphenyl group (two benzene rings with a carbon-to-carbon bond between carbon 1 on the first ring and carbon 1'

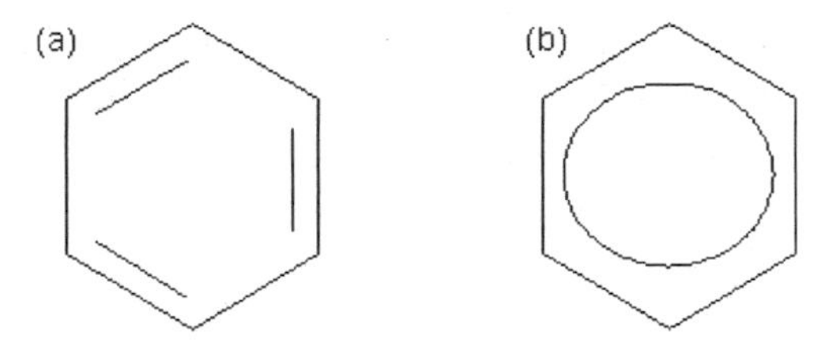

FIGURE 1-1 Benzene ring (a) with conjugated bonds and (b) with inner ring depicting conjugated bonds.

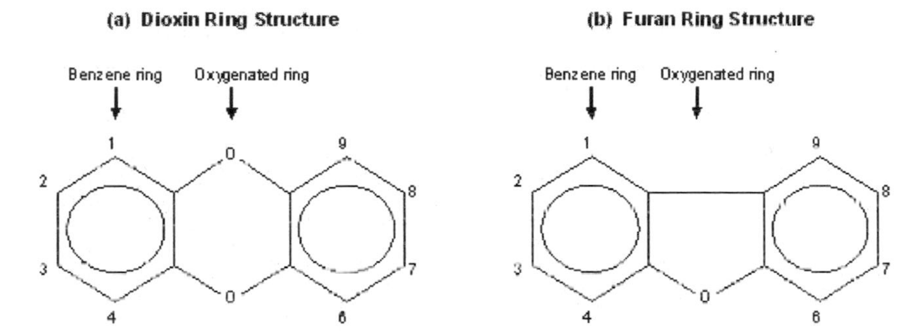

FIGURE 1-2 Double benzene ring structures of (a) dioxins and (b) furans.

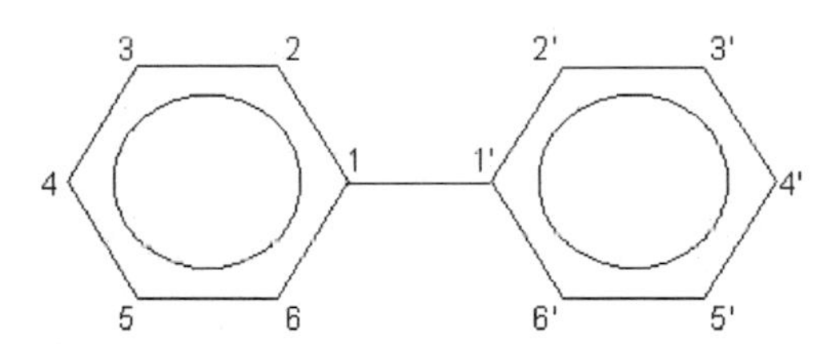

FIGURE 1-3 Biphenyl ring structure of PCBs.

on the second ring) (Figure 1-3). Examples of some PCDDs, PCDFs, and PCBs of interest are shown in Figure 1-4. Each chemical compound from any of these subclasses is referred to as a congener. Brominated or mixed halogenated congeners within these classes of compounds or within other chemical classes, such as the polyhalogenated naphthalenes, benzenes, azobenzenes, and azoxybenzenes, have not been evaluated as extensively and are not addressed in the Reassessment. TCDD, the most studied and one of the most toxic members of these classes of compounds, is the designated reference chemical for the Reassessment and for other related litera-

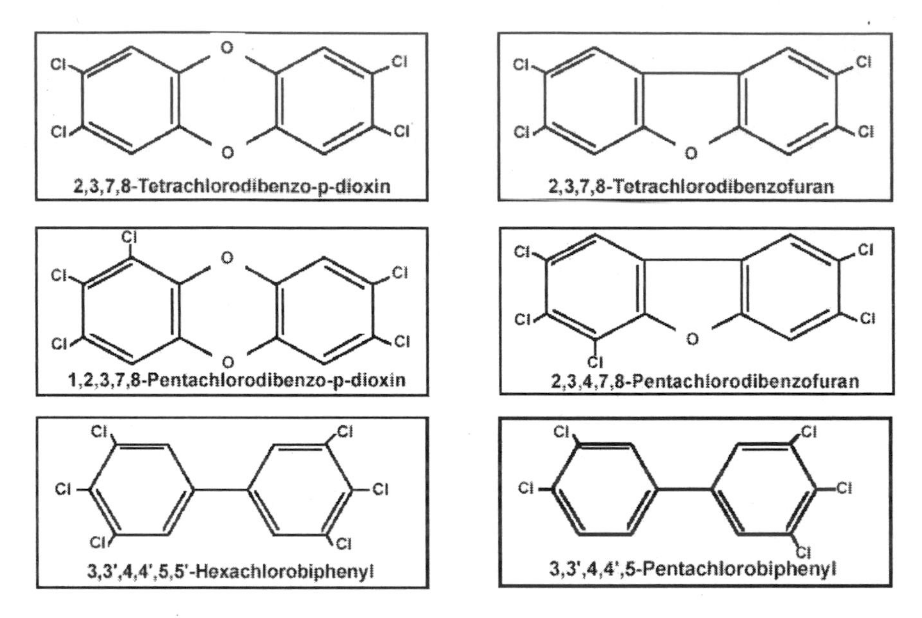

FIGURE 1-4 Examples of toxic PCDDs, PCDFs, and PCBs of interest in the Reassessment.

ture. PCDDs and PCDFs are tricyclic aromatic compounds with similar physical and chemical properties—properties shared by specifically configured, or coplanar (a flat configuration), dioxin-like PCBs. The Reassessment uses the terms "dioxins" and "dioxin-like compounds" in reference to any individual or any mixture of the addressed chemicals. These are general terms that describe chemicals that share defined similarities, including chemical structure and biological and toxicological character. However, of the several hundred HAH congeners, only 29 are considered to have significant toxicity and to induce a common battery of toxic responses through similar biological modes of action. The evaluation of dioxin-like congeners within the Reassessment focuses on dioxins and DLCs with TCDD-like toxicity and those generally considered the most associated with environmental and human health risks. These chemicals include PCDDs and PCDFs that retain chlorine substitutions at positions 2, 3, 7, and 8 on the benzene rings (see Figure 1-2). The remaining evaluated TCDD-like congeners include the PCBs with four or more chlorines in the lateral positions (3, 3', 4, 4', 5, or 5'), with established TCDD-like environmental and biological behaviors, and particularly the mono- and non-ortho PCBs—that is, PCBs with one or no, respectively, chlorine substitution in the ortho position (2, 2', 6, or 6') on the benzene rings (see Figure 1-3). Studies show that the TCDD-like toxicity of the PCB congener increases with a larger number of chlorines in the lateral positions and one or no chlorines in the ortho position. Also, when a PCB has only one or no substitution in the ortho position, the atoms of the PCB congener can line up in a coplanar or flat configuration, making these the most toxic of the TCDD-like PCBs. Evaluation of the chemical congeners addressed in the Reassessment is considered sufficient to characterize environmental chlorinated dioxins (Reassessment, Part I, p. 1-5, lines 4 to 6).

Experimental evidence indicates that TCDD acts by way of binding to an intracellular protein, the aromatic hydrocarbon receptor (AHR), a ligand-dependent transcription factor that functions in partnership with a second protein, the AHR nuclear translocator protein (ARNT) to stimulate alterations in gene expression that result in toxic and biological effects. AHR is present throughout the animal kingdom, including invertebrates like the fruit fly and the clam. In addition to AHR binding, several other molecular events are necessary for AHR-dependent biological and toxic effects to occur, and there are significant species differences in those events, so quantitative cross-species comparisons based only on AHR binding may not provide accurate or dependable information about their AHR responsiveness or the possible AHR dependent responses. The invertebrate AHR does not bind xenobiotic ligands (that is, TCDD, other dioxins, and DLCs), and it is not associated with toxic end points, suggesting that the role of AHR as a mediator of toxic or adaptive responses might be a function acquired during vertebrate evolution and superimposed on an endogenous physiological role.

TOXIC EQUIVALENCY FACTORS

TCDD, other dioxins, and DLCs are generally present as complex mixtures in environmental, food, and biological matrices, including humans and other animals. To address the complexity of risk assessment for TCDD-related compounds, the concept of TEFs is used in the Reassessment. The TEF concept originated as a method for evaluating health risks associated with closely related chemicals with similar mechanisms of action but different potencies. The criteria for including a chemical in the TEF concept are described in detail in the Reassessment, Part II, Chapter 9. The TEF of each chemical congener is determined by evaluating available congener-specific data (primarily in vivo data), and the congener is then assigned an "order of magnitude" estimate of relative toxicity compared with the prototypical and most potent HAH, 2,3,7,8-TCDD. By using those factors, the toxicity of a mixture is expressed in terms of its total toxic equivalent quotient (TEQ), which is the amount of TCDD that it would take to equal the combined toxic effect of all contributing congeners within the mixture. The TEF value of each congener within a mixture is multiplied by its concentration, and the products (TEQs) are summed to yield the total TEQ of the mixture, which is the estimate of the total toxicity of the mixture.

In 1997, a team of experts convened by the WHO European Center for Environment and Health and the International Program on Chemical Safety (IPCS) evaluated a large database of experimental data of the relative potencies for PCDDs, PCDFs, and dioxin-like PCBs to establish consensus TEF values for these compounds in mammals, birds, and fish. Human and mammalian TEFs were constructed by an approach that gave more weight to in vivo toxicity data than to in vitro data. Moreover, among the different in vivo studies available for establishing TEFs, the basis for selecting the most relevant in vivo toxicity study was the length of exposure, with chronic exposures ranking highest and acute exposures lowest in relevance. The team concluded that an additive TEF model served as the most feasible risk assessment method for complex mixtures of TCDD-like PCDDs, PCDFs, and PCBs. Although several TEF/TEQ schemes exist for TCDD-related compounds, the Reassessment recommends using the international WHO TEF scheme of values, proposed and published by WHO (IPCS 1998a), to assign toxic equivalency for the Reassessment. Table 1-1 presents WHO TEFs established for humans and mammals for 7 PCDDs, 10 PCDFs, and 12 dioxin-like PCBs. TEF assignments continue to evolve in accordance with emerging science and iteration, and WHO recommended revisiting TEF values every 5 years, with review in 2005. For additional information on the TEF/TEQ approach, see Chapter 3 of the Reassessment, Part II.

To facilitate evaluation of human health risks and regulatory control of exposure to mixtures of TCDD, other dioxins, and DLCs, EPA, using all available data, has incorporated the TEF concept and method into the risk assessment process since 1987. The Reassessment considers the application, limitations, and uncertainties when using TEFs. Part II, Chapter 9, of the Reassessment describes the application of the TEF method for TCDD, other dioxins, and DLCs and addresses the uncertainties in detail. Overall, the use of the TEF method is currently the most reliable and best evaluated approach for assessing the potential toxic potency of complex mixtures of DLCs. TEFs/TEQs are addressed in further detail in Chapter 3.

EXPOSURE CHARACTERIZATION

EPA classifies sources of TCDD, other dioxins, and DLCs into five categories—(1) combustion; (2) metal smelting, refining, and processing; (3) chemical manufacturing and processing; (4) biological and photochemical processes; and (5) reservoir sources. Combustion sources include incineration of various types of waste (municipal solid, sewage sludge, medical, and hazardous), burning of fuels (coal, wood, and petroleum products),

TABLE 1-1 TEFs for Humans and Nonhuman Mammals

PCDD Congeners	WHO TEF
2,3,7,8-TCDD	1
1,2,3,7,8-PeCDD	1
1,2,3,4,7,8-HxCDD	0.1
1,2,3,7,8,9-HxCDD	0.1
1,2,3,6,7,8-HxCDD	0.1
1,2,3,4,6,7,8-HpCDD	0.01
1,2,3,4,6,7,8,9-OCDD	0.0001

PCDF Congeners	WHO TEF
2,3,7,8-TCDF	0.1
1,2,3,7,8-PeCDF	0.05
2,3,4,7,8-PeCDF	0.5
1,2,3,4,7,8-HxCDF	0.1
1,2,3,7,8,9-HxCDF	0.1
1,2,3,6,7,8-HxCDF	0.1
2,3,4,6,7,8-HxCDF	0.1
1,2,3,4,6,7,8-HpCDF	0.01
1,2,3,4,7,8,9-HpCDF	0.01
1,2,3,4,6,7,8,9-OCDF	0.0001

PCB Congeners		WHO TEF
IUPAC Number	Structure	
77	3,3',4,4' TCB	0.0001
81	3,4,4',5-TCB	0.0001
105	2,3,3',4,4'-PeCB	0.0001
114	2,3,4,4',5-PeCB	0.0005
118	2,3',4,4',5-PeCB	0.0001
123	2',3,4,4',5-PeCB	0.0001
126	3,3',4,4',5-PeCB	0.1
156	2,3,3',4,4',5-HxCB	0.0005
157	2,3,3',4,4',5'-HxCB	0.0005
167	2,3',4,4',5,5'-HxCB	0.00001
169	3,3',4,4',5,5'-HxCB	0.01
189	2,3,3',4,4',5,5'-HpCB	0.0001

Abbreviations: PeCDD, pentachlorodibenzo-*p*-dioxin; HxCDD, hexachlorodibenzo-*p*-dioxin; HpCDD, heptachlorodibenzo-*p*-dioxin; OCDD, octachlorodibenzo-*p*-dioxin; TCDF, tetrachlorodibenzofuran; PeCDF, pentachlorodibenzofuran; HxCDF, hexachlorodibenzofuran; HpCDF, heptachlorodibenzofuran; OCDF, octachlorodibenzofuran; TCB, tetrachlorobiphenyl; PeCB, pentachlorobiphenyl; HxCB, hexachlorobiphenyal; HpCB, heptachlorobiphenyl.
SOURCE: IPCS 1998a. Reprinted with permission; copyright 1998, World Health Organization.

forest fires and open burning of waste materials, and high-temperature processes (e.g., cement kiln operations). Combustion sources produce PCDDs, PCDFs, and limited amounts of PCBs (commercially manufactured in large quantities from about 1930 until 1977). Metallurgical operations (e.g., iron ore sintering, steel production, and scrap metal recovery) can produce PCDDs and PCDFs, which are also formed as by-products of chemical processing (e.g., manufacture of chlorine-bleached wood pulp and phenoxy herbicides). PCDDs and PCDFs can also be formed under such environmental conditions as composting via microorganism action on chlorinated phenolic compounds. Studies have also reported that these chemicals form during photolysis of highly chlorinated phenols, such as pentachlorophenol, although it has been demonstrated only under laboratory conditions. Four of the source categories (combustion, metallurgical processing, chemical manufacturing and processing, and biological and photochemical processes) are collectively referred to as contemporary formation sources. In contrast, reservoir sources are not considered in the quantitative inventory of contemporary formation sources because they involve the recirculation of previously formed compounds that have already partitioned into air, water, soil, sediment, and biota. However, the Reassessment recognizes that the contribution of reservoir sources to human exposure may be significant, perhaps contributing half or more of total background TEQ exposure. For any given time period, releases from both contemporary formation sources and from reservoir sources determine the overall amount of TCDD, other dioxins, and DLCs released to the accessible environment.

The Reassessment gives an inventory of environmental releases of PCDDs, PCDFs, and TCDD-like PCBs for the United States based on two reference years, 1987 and 1995. An updated inventory for reference year 2000 was published in 2005 and was included in the committee's review. EPA's best estimate of releases of these compounds to air, water, and land from reasonably quantifiable sources in 2000 was approximately 1,500 g TEQ_{DF}-WHO ([DF] dioxins and furans), representing an 89% decrease from a 1987 best estimate of 14,000 g TEQ_{DF}-WHO. U.S. environmental releases of PCDDs and PCDFs occur from an expansive variety of sources but are dominated by releases to the air from combustion sources. The decrease in estimated releases of PCDDs and PCDFs from 1987 to 2000 is largely attributed to reductions in air emissions from municipal and medical waste incinerators; further reductions are anticipated. Three types of combustion sources contributed approximately 70% of all quantifiable environmental releases in 1995: municipal waste incinerators, backyard burning of refuse in barrels, and medical waste incinerators, representing 38%, 19%, and 14% of total environmental releases, respectively. A number of investigators have proposed that the U.S. inventory underestimates releases from contemporary formation sources partly because of the lack of sufficient data from sources that can emit TCDD and related compounds, such as land fires; unquantifiable or poorly quantifiable sources, such as agricultural burning; and the possibility of unknown sources. Additional observations in the Reassessment regarding sources are concerns about insufficient data or estimates from nonpoint sources (e.g., urban stormwater runoff and rural soil erosion) and the likelihood that total nonpoint-source releases are substantially larger than point-source releases. Evidence also indicates that current emissions of TCDD, other dioxins, and DLCs to the U.S. environment result principally from anthropogenic activities, as supported by correlations in the rise in environmental levels of these compounds and a period of rapid increase in industrial activities, lack of significant natural sources, and observations of higher body burdens in industrialized versus less industrialized countries. PCDDs, PCDFs, and PCBs share similar properties, including lipophilicity, hydrophobicity, and resistance to degradation. Consequently, these intrinsically stable compounds are found throughout the world in practically all environmental media, including air, water, soil, sediment, food, and food products. The amount of time required for a chemical to lose one-half of its original concentration, known as its half-life, varies by substance. The chemical half-lives of mixtures change with time, as the shorter-lived substances disappear and the proportion with longer half-lives increases. (For further discussion on chemical half-lives, see commentary (Part II, Volume 2, Chapter 2).

The Reassessment defines background exposure to TCDD, other dioxins, and DLCs as exposure that would occur in an area without known point sources of the contaminants. Background exposure includes exposure via the commercial food supply, air, or soil but not any significant occupational exposure. Background exposure estimates are based on the monitoring of data from environmental sites and other media void of known contaminant sources and on pharmacokinetic models using body burden data from nonoccupationally exposed populations. High concentrations, measured in parts per trillion (ppt) and higher, are found in soil, sediments, and biota because of their recalcitrant nature and their physical-chemical properties. Low concentrations, measured in parts per quadrillion (ppq) and picograms per cubic meter (pg/m^3), are found in water and air, respectively.

Estimates for background concentrations of DLCs in environmental media and in food are based on studies conducted at various locations in North America. The number of locations examined for environmental media estimates in those studies was small, and it is not known whether the estimates adequately reflect the full range of variation across the United States. Food estimates were derived from statistically based national surveys, nationwide sampling networks, food fat concentration measurements, samples collected from retail stores, and samples obtained from biohabitat.

PCDD, PCDF, and PCB TEQ-WHO concentrations in environmental

media and food are presented in Table 1-2. Measurable quantities of these in environmental media and food in the United States were found to be similar to quantities measured in Europe. Evidence from Europe suggests a decline in dioxin and furan concentrations in food products during the 1990s. Although no systematic study of temporal trends in dioxin concentrations in food has been conducted in the United States, at least one study determined that meat now contains lower concentrations of TCDD, other dioxins, and DLCs than samples from the 1950s through the 1970s contained. The U.S. Department of Agriculture is conducting a nationwide survey of dioxin concentrations in beef, pork, and poultry that should allow for a time-trend analysis.

The average PCDD, PCDF, and PCB tissue concentration for the general adult U.S. population in the late 1990s, based on EPA's estimate, was 25 ppt TEQ_{DFP}-WHO ([DFP] dioxins, furans, and PCBs), lipid basis (Reassessment, Part III, p. 4-15). This estimate suggests average tissue concentrations have declined from the estimated 55 ppt in the late 1980s and early 1990s. Because new emissions of TCDD, other dioxins, and DLCs have been declining since the 1970s, it is reasonable to expect that concentrations in food, human diet, and, ultimately, human tissue have also declined during this time.

The Reassessment acknowledges that characterization of national background concentrations of TCDD, other dioxins, and DLCs in tissue is uncertain because current data are not statistically representative of general populations. Also, tissue concentrations are a function of age and year of birth.

HEALTH EFFECTS

On a global scale, exposure to TCDD, other dioxins, and DLCs resulting from accidental, occupational, or incidental exposure through dermal contact, inhalation, or ingestion has been associated with adverse effects on human health. In the early 1900s, workers involved in distilling, processing, or producing chlorine-based chemicals presented with symptoms characteristic of those currently associated with TCDD poisoning, including severe cases of chloracne and various degrees of fatigue. Soil contaminated with TCDD at 300 ppb caused the 1983 evacuation of the town of Times Beach, Missouri, and allegedly was responsible for the deaths of local animals and for a variety of human and animal illnesses. In a January 2003 press release, the Institute of Medicine announced that reexamination of six studies of herbicide-exposed veterans revealed sufficient evidence of an association between herbicide defoliants, or their contaminants, which included TCDD, sprayed by U.S. forces in Vietnam and the risk of developing chloracne, chronic lymphocytic leukemia, and soft tissue sarcoma.

Some studies suggest that exposure of chemical workers to very high concentrations of TCDD, other dioxins, and/or PCBs (body burdens of 100-1,000 times background) are associated with an increased incidence of cancer (Flesch- Janys et al. 1995; Hooiveld et al. 1998; Steenland et al. 1999). Other studies of highly exposed populations suggest that TCDD, other dioxins, and PCBs can have reproductive and developmental effects (Eskenazi et al. 2000; Kogevinas 2001; Revich 2002; Vreugdenhill et al. 2002a; Pesatori et al. 2003). The long-term effects of low-level exposure to TCDD, other dioxins, or DLCs normally experienced by the general population are not known, nor is the clinical significance of biochemical biomarkers, such as enzyme induction at or near background-level exposures. Focal points of research include organ and organ-system effects and elucidation of the cellular mechanisms through which these effects occur.

COMMITTEE CHARGE AND RESPONSE

In May 2004, EPA asked the NRC to review the revised draft reassessment titled *Exposure and Human Health Reassessment of 2,3,7,8-Tetrachlorodibenzo-p-dioxin (TCDD) and Related Compounds* (2003 version, publicly released in October 2004) and to assess whether EPA's risk estimates are scientifically robust and whether there is a clear delineation of all substantial uncertainties and variability (see Box 1-1 for the complete statement of task). In response, the NRC formed the Committee on EPA's Exposure and Human Health Reassessment of TCDD and Related Compounds, a panel of 18 members that included experts in exposure assessment; food exposure pathways; pharmacokinetics; physiologically based pharmacokinetic modeling; benchmark dose modeling; dose-response modeling; molecular and cellular aspects of receptor-mediated responses; toxicology with specialties in cancer, reproduction, development, and immunology; epidemiology; reproductive physiology and medicine; pediatric biology and medicine; statistics; risk assessment (both qualitative and quantitative); and uncertainty analysis (see Appendix A for details).

The committee held three public meetings in Washington, DC, to collect information, meet with researchers and decision makers, and accept testimony from the public. The committee met two additional times, in executive session, to complete its report. Although the committee reviewed all three parts of the Reassessment, it focused primarily on Part III—Dioxin: Integrated Summary and Risk Characterization for 2,3,7,8-Tetrachlorodibenzo-*p*-Dioxin (TCDD) and Related Compounds, as directed by the study charge. The committee also considered new peer-reviewed studies published since Part III of the Reassessment was last revised and before the committee held its final meeting in July 2005. However, because the committee was charged to review EPA's Reassessment, conducting a compre-

TABLE 1-2 Summary of North American PCDD, PCDF, and PCB TEQ WHO Concentrations in Environmental Media and Food (Whole Weight Basis)

Media	PCDDs, PCDFs[a]	References[b]	PCBs[a]	References[b]	Mean Total PCDD, PCDF, PCBs
Urban soil, ppt	n = 270 9.3 ± 10.2[c] Range = 2 to 21	EPA 1985, 1996a, 2000a; Nestrick et al. 1986; Birmingham 1990; Pearson et al. 1990 ; NIH 1995; Rogowski et al. 1999	n = 99 2.3	EPA 2000a	11.6
Rural soil, ppt	n = 354 2.7[c] Range = 0.11 to 5.7	EPA 1985, 1996a, 2000a; Birmingham 1990; Pearson et al. 1990; Reed et al. 1990; MRI 1992; van Oostdam and Ward 1995; Tewhey Assoc. 1997; Rogowski et al. 1999; Rogowski and Yake 1999	n = 62 0.59	EPA 2000a	3.3
Sediment, ppt	n = 11 5.3 ± 5.8[c] Range = <1 to 20	Cleverly et al. 1996	n = 11 0.53 ± 0.69[c]	Cleverly et al. 1996	5.8
Urban air, pg/m^3	n = 106 0.12 ± 0.094[c] Range = 0.03 to 0.2	CDEP 1988, 1995; Hunt et al. 1990; Hunt and Maisel 1990; Maisel and Hunt 1990; OHEPA 1995; Smith et al. 1989, 1990	n = 53 0.0009[d]	Hoff et al. 1992	0.12

Rural air, pg/m^3	n = 60 0.013[c] Range = 0.004 to 0.02	CDEP 1995; OHEPA 1995; Cleverly et al. 2000	n = 53 0.00071	Cleverly et al. 2000	0.014
Freshwater fish and shellfish, ppt	n = 222 1.0[c]	EPA 1992; Fiedler et al. 1997; Jensen et al. 2000; Jensen and Bolger 2001	n = 1 composite of 10 samples plus 6 composites of 1.2[e,f]	Mes and Weber 1989; Mes et al. 1991, Schecter et al. 1997	2.2
Marine fish and shellfish, ppt	n = 158 0.26[c]	Fiedler et al. 1997; Jensen et al. 2000	n = 1 composite of 13 samples plus 6 composites 0.25[e,f]	Mes et al. 1991; Schecter et al. 1997	0.57
Water, ppq	n = 236 0.00056 ± 0.00079	Meyer et al. 1989 ; Jobb et al. 1990	—[g]	—	0.00056
Milk, ppt	n = 8 composites 0.018[f]	Lorber et al. 1998	n = 8 composites 0.0088	Lorber et al. 1998	0.027
Dairy, ppt	n = 8 composites 0.12[f]	Based on data from Lorber et al. 1998	n = 8 composites 0.058	Based on data from Lorber et al. 1998	0.18
Eggs, ppt	n = 15 composites 0.081[f]	Hayward and Bolger 2000	n = 18 plus 6 composites of 0.10[e,f]	Mes and Weber 1989; Mes et al. 1991; Schecter et. 1997	0.13
Beef, ppt	n = 63 0.18 ± 0.11 Range = 0.11 to 0.95	Winters et al. 1996a	n = 63 0.084	Winters et al. 1996b	0.26

continued

TABLE 1-2 Continued

Media	PCDDs, PCDFs[a]	References[b]	PCBs[a]	References[b]	Mean Total PCDD, PCDF, PCBs
Pork, ppt	n = 78 0.28 ± 0.28 Range = 0.15 to 1.8	Lorber et al. 1997	n = 78 0.012	Lorber et al. 1997	0.29
Poultry, ppt	n = 78 0.068 ± 0.070 Range 0.03 to 0.43	Ferrario et al. 1997	n = 78 0.026	Ferrario et al. 1997	0.094
Vegetable fats, ppt	n = 30 0.056 ± 0.24[h]	Versar 1996	n = 5 composites 0.037[f]	Mes et al. 1991	0.093

[a]Values are the arithmetic mean TEQs, in ppt, and standard deviations. Nondetects were set to one-half the limit of detection, except for soil, PCDDs, and PCDFs in vegetable fats for which nondetects were set to zero.
[b]A full list of references is found in Part I, Vol. 2, Chapter 3 of the Reassessment NAS review draft (December 2003).
[c]The values for environmental media are means of the data but lack the spatial representativeness to be considered true national means.
[d]Based on data from Canadian air, as reported by Hoff et al. (1992). Not used in U.S. background exposure estimates in Part I, Vol. 2, Chapter 4 of the Reassessment NAS review draft (December 2003).
[e]The values for fish lack the statistical significance to be considered true means; the values of the other food groups were derived from statistically based surveys and can be considered true national means. The PCCD and PCDF concentrations are species-specific ingestion-weighted average values.
[f]Standard deviations could not be calculated because of limitations of the data (composite analyses).
[g]Congener-specific PCB data are sparse.
[h]TEQ calculated from Versar (1996) by setting nondetects to zero.
SOURCE: EPA 2003a.

BOX 1-1 Statement of Task

The National Academies' National Research Council will convene an expert committee that will review EPA's 2003 draft reassessment of the risks of dioxins and dioxin-like compounds to assess whether EPA's risk estimates are scientifically robust and whether there is a clear delineation of all substantial uncertainties and variability. To the extent possible, the review will focus on EPA's modeling assumptions, including those associated with the dose-response curve and points of departure; dose ranges and associated likelihood estimates for identified human health outcomes; EPA's quantitative uncertainty analysis; EPA's selection of studies as a basis for its assessments; and gaps in scientific knowledge. The study will also address the following aspects of the EPA reassessment: (1) the scientific evidence for classifying dioxin as a human carcinogen; and (2) the validity of the non-threshold linear dose-response model and the cancer slope factor calculated by EPA through the use of this model. The committee will also provide scientific judgment regarding the usefulness of toxicity equivalence factors (TEFs) in the risk assessment of complex mixtures of dioxins and the uncertainties associated with the use of TEFs. The committee will also review the uncertainty associated with the reassessment's approach regarding the analysis of food sampling and human dietary intake data, and, therefore, human exposures, taking into consideration the Institute of Medicine's report *Dioxin and Dioxin-Like Compounds in the Food Supply: Strategies to Decrease Exposure*. The committee will focus particularly on the risk characterization section of EPA's reassessment report and will endeavor to make the uncertainties in such risk assessments more fully understood by decision makers. The committee will review the breadth of the uncertainty and variability associated with risk assessment decisions and numerical choices, for example, modeling assumptions, including those associated with the dose-response curve and points of departure. The committee will also review quantitative uncertainty analyses, as feasible and appropriate. The committee will identify gaps in scientific knowledge that are critical to understanding dioxin reassessment.

hensive and thorough review of all TCDD-related materials published since 2003, reassessing TEF values, and re-creating the risk assessment were outside of the scope of the statement of task.

The present report is the product of the efforts of the entire NRC committee and underwent extensive, independent, external review overseen by the NRC's Report Review Committee. It specifically addresses and is limited to the statement of task as agreed upon by the NRC and EPA.

The remaining chapters of this report comprise the findings of the Committee on EPA's Exposure and Human Health Reassessment of TCDD and Related Compounds. Chapter 2 provides conceptual text on how to address variability and uncertainty in risk assessment. Chapter 3 evaluates the usefulness and uncertainties of TEFs in the risk assessment of complex mixtures of TCDD, other dioxins, and DLCs and discusses various approaches to dose metrics. Chapter 4 addresses exposure characterization in terms of sources, environmental fate, environmental media concentrations, food concentrations, background exposures, and potentially highly exposed populations. Chapter 5 reviews EPA's assessment of the carcinogenicity of TCDD other TCDD, other dioxins, and DLCs, including the qualitative characterization of their carcinogenicity, the validity of the nonthreshold linear dose-response model, and the use of the animal bioassay and epidemiological data to quantify the dose response. Chapter 6 reviews EPA's assessment of noncancer end points, including immune function, reproduction, and development. Chapter 7 focuses on risk characterization. Chapter 8 summarizes the committee's conclusions and recommendations and succinctly addresses each component of the statement of task.

2

General Considerations of Uncertainty and Variability, Selection of Dose Metric, and Dose-Response Modeling

Health risk assessments now typically include discussion of variability (real differences) and uncertainty (fundamental lack of knowledge) and often use probabilistic risk assessment methods to characterize variability and uncertainty in the estimates of risks. Prior National Research Council (NRC) reports and U.S. Environmental Protection Agency (EPA) documents make clear the need for these characterizations; for example, they emphasize that

> uncertainty forces decision-makers to judge how probable it is that risks will be overestimated or underestimated for every member of the exposed population, whereas variability forces them to cope with the certainty that different individuals will be subjected to risks both above and below any reference point one chooses (NRC 1994, p. 237)

and that

> [i]n successive versions of its cancer guidelines, EPA expressed increasing emphasis on a full examination of uncertainties, with the recognition that both qualitative and quantitative approaches to uncertainty assessment are important and can (applied appropriately) help clarify the nature of assessment findings. The use of sophisticated uncertainty tools also involves substantial issues of science and mathematics, as well as specialized issues such as the appropriate presentation and characterization of probabilistic estimates in the decision making context where appropriate. (EPA 2004a, p. 49)

Significant uncertainties remain in understanding human health risks from 2,3,7,8-tetrachlorodibenzo-*p*-dioxin (TCDD), other dioxins, and dioxin-like compounds (DLCs), in spite of very large investments in data collection and research.

Variability among members of the population is an important consideration in understanding risks. Variability results from the wide range of environmental sources and human interactions with them, as well as from physiological and genetic differences that might influence the relative susceptibility of humans and other species to adverse health effects from exposure. For example, sources of variability associated with human health outcomes include the inherent genetic diversity of human populations, which currently remain difficult to address quantitatively. Abundant evidence demonstrates complex gene-environment interactions for many complex human diseases, immune system dysfunction, and other disorders in which TCDD, other dioxins, and DLCs might be implicated.

Adding more complexity, the risks from TCDD, other dioxins and DLCs continue to change over time because of changing exposures, and understanding of the risks continues to evolve with the collection of more data. Any assessment reflects the snapshot of the information available at that time, and analysts should recognize that additional information might later reveal evidence that differs from prior assumptions.

One of the charges to the committee emphasized reviewing the Reassessment[1] "to assess whether EPA's risk estimates are scientifically robust and whether there is a clear delineation of all substantial uncertainties and variability." Risk assessment in the case of TCDD, other dioxins, and DLCs represents a formidable task because of the size of the available database and the complexity of numerous issues. EPA collated and presented a massive database on TCDD, other dioxins, and DLCs, on which the committee commented specifically in the chapters that follow. This chapter identifies the major categories of decisions that analysts generally make when developing risk estimates in the context of the four traditional steps of risk assessment: hazard identification and classification, exposure assessment, dose-response assessment, and risk characterization (NRC 1983). The Reassessment deals with complexities in the risk assessment of TCDD, other dioxins and DLCs by making specific choices as described in this chapter, but EPA could alternatively use a probabilistic approach. Typically, risk assessments should address uncertainties that derive from conceptualizations and fundamental choices among competing options in a way that clearly identifies the quantitative impacts of alternatives. When there are two or more plausible interpretations, a risk assessment should make clear

[1]*The Exposure and Human Health Reassessment of 2,3,7,8-Tetrachlorodibenzo-p-dioxin (TCDD) and Related Compounds* (EPA 2003a, Part I; 2003b, Part II; 2003c, Part III) is collectively referred to as the Reassessment.

that such alternatives give rise to uncertainty. To this end, a risk assessment should identify the key uncertainties (those that drive the risk estimates) and make clear how selection of specific alternative assumptions influences the risk assessment results.

In general, the choice of individual or population risk metric that is modeled influences the appropriate characterization of variability and uncertainty in risk (Thompson and Graham 1996). The Reassessment strives to present a comprehensive baseline risk assessment intended to cover all potential sources. This generic approach results in limited discussions of variability and uncertainty. The committee found that the lack of a specific context and absence of a focused exposure assessment that would link sources to potential health effects in individuals, or in the population, severely limited both EPA's and the committee's abilities to appropriately characterize variability and uncertainty in risk estimates related to exposure to TCDD, other dioxins, and DLCs.

HAZARD CLASSIFICATION

In the context of the Reassessment, EPA faced the decisions of assigning a hazard classification for TCDD, and for other dioxins and DLCs, including mixtures. Hazard classification typically focuses on characterizing the weight of the evidence with respect to potential health effects. For cancer risk, the cancer guidelines (EPA 2005a, also see Appendix B) outline specific criteria for classifying substances into the following categories:

1. Carcinogenic to humans
2. Likely to be carcinogenic to humans
3. Suggestive evidence of carcinogenic potential
4. Inadequate evidence to assess carcinogenic potential
5. Not likely to be carcinogenic to humans

The charge to the committee stated that it should address "the scientific evidence for classifying dioxin as a human carcinogen."

The committee believes that the scientific evidence on cancer causation usually falls within a continuum, and classification often artificially places apparent bright lines (e.g., in distinguishing a "known human carcinogen" from a "likely human carcinogen"). In Chapter 5, the committee reviews and comments on EPA's decisions with respect to its determinations of cancer classification.

With respect to noncancer end points, the committee notes that EPA does not use a rigorous approach for evaluating evidence from studies and the weight of their evidence in the Reassessment. The committee finds that EPA's lack of systematic evaluation and classification of the noncancer evidence left significant ambiguity about the basis for some of EPA's decisions implied in the report (e.g., the decision not to identify a critical effect or to develop a reference dose [RfD]). The Reassessment provides an extensive catalog of studies but does not synthesize the significant insights or provide clear assessments of the key uncertainties in a way that allows the reader to determine the impact of various choices made.

In general, the use of a rigorous evaluation process for noncancer hazards would lead to improved characterization of noncancer risks. In the context of the Reassessment and any future iterations of this analysis, the committee suggests that EPA focus its efforts on improving its quantitative characterization of the risks, including noncancer risks, and not devote substantial effort to further carcinogen classification for TCDD, other dioxins, and DLCs, as discussed in Chapter 5.

EXPOSURE ASSESSMENT

EPA provided the committee with an updated exposure inventory (EPA 2005b), which provides an extensive review of the existing database of exposure data for TCDD, other dioxins, and DLCs. The review also provides a useful qualitative review of the level of confidence in the data for various sources, although the Reassessment does not quantitatively characterize the uncertainty associated with low-confidence data. Although the Reassessment (Part III, p. 4-6) specifically mentions the possibility of unknown sources causing underestimation of releases from contemporary sources, it does not attempt to correct the incomplete accounting of sources in historical data or adjust current data to address anticipated discoveries of other sources. Thus, EPA implicitly assumed that the exposure assessment sufficiently captures the exposure sources so that any additional new sources identified would not significantly alter its estimates. The committee discusses this choice in more detail in Chapter 4 and suggests additional analyses that might further explore the impacts of this assumption.

The updated exposure inventory devotes considerable attention to documenting how the nature and magnitude of dominant exposure sources changed over time. The substantial amount of new evidence of significant declines in measured concentrations of TCDD, other dioxins, and DLCs over the past several decades reflects EPA's specific management efforts targeted at reducing exposure from some sources (e.g., pulp and paper mills, medical and municipal waste incineration, and ball clay[2]). Referring

[2]The term ball clay originated from an early English mining practice of rolling the highly plastic clay into balls weighing 30 to 50 lb. Ball-clay uses historically included serving as a supplement in animal feeds (as in chicken feed). In 1996, as a result of investigations into the source of contamination with TCDD and other dioxins in chicken fat, investigators measured relatively high levels of TCDD and related compounds in ball clay (FDA 1997).

specifically to TCDD, EPA notes that "dioxin levels in the environment have been declining since the 1970s..., and it is reasonable to expect that levels in food, human intake, and ultimately, human tissue have also declined over this period. The changes in tissue levels are likely to lag the decline seen in environmental levels, and the changes in tissue levels cannot be assumed to occur proportionally with declines in environmental levels" (Reassessment, Part III, p. 4-16). Changing concentrations in the environment over time provides another substantial uncertainty in risk assessment, because EPA must decide whether to use specific "snapshot in time" concentrations for risk assessment or whether to extrapolate or average such changing concentrations over time. Given the timing of the updated exposure inventory, it was not clear to the committee how EPA intends to use the updated inventory information in the context of estimating current exposures.

Another area of uncertainty lies in determining what constitutes background exposures in the general population. EPA carefully defines "background" in a prominent footnote (Reassessment, Part III, p. 1-1), and the committee concurs that this approach is appropriate and is clearly presented in the Reassessment. However, the uncertainty associated with potential discoveries of "new sources" will remain an issue that EPA may need to analyze further. For example, the Reassessment added a chapter on ball clays in the latest iteration.

Yet another area of uncertainty is determination of background levels when many samples lie below the analytical limit of detection. This issue arises in any exposure assessment, and several widely used options address it (e.g., assume all nondetects are true zeroes, assign a value of either ½ or 1 times the detection limit, or fit a distribution to the data). The committee noted that EPA did not pick a single consistent approach (see the note to the summary table at the bottom of Part III, p. 4-32) or provide a clear quantitative indication of the importance of the choice of strategy for dealing with nondetects, which creates inconsistencies in the Reassessment. The committee recommends that EPA clearly and quantitatively explore how different strategies for dealing with nondetects affect exposure assessment results, as discussed in Chapter 4. If these alternative approaches produce very different results, then EPA should further consider the implications of specific options.

Another major source of uncertainty stems from the selection of a dose metric. The Reassessment could provide exposure estimates for a wide range of dose metrics and averaging times to support the spectrum of possible dose-response assessment choices. This important issue is discussed in more detail below. The Reassessment also provides little insight about bioavailability, an issue that frequently falls between the domains of the exposure assessment and dose-response assessment.

Finally, the Reassessment provides very little information about the amount of individual variability in exposure. EPA describes how average daily toxic equivalent quotient (TEQ) varies as a function of age (Reassessment, Part III, pp. 4-16, 4-17, and 4-35), although it does not provide a measure of the variability around these estimates (that is, the population distribution of exposures within each age group). EPA's description may implicitly give the impression of very limited variability within the population, which may not be the case. However, the Reassessment provides some good examples of other parameters that may influence interindividual variability. For example, considering the variability in total fat consumption, the Reassessment suggests that TCDD intakes in the general population could extend to levels at least three times higher than the mean (Reassessment, Part III, p. 4-19). The exposure assessment also demonstrates that TCDD intake for children based on age-specific food consumption and average food concentrations exceeds adult intake estimates on a body-weight basis (although their intake on a mass basis is lower) (Reassessment, Part III, p. 4-35). These examples also illustrate the difficulties that arise in choosing an appropriate overall averaging time for exposure.

ASSESSMENT OF OTHER DIOXINS AND DLCS

The challenge of characterizing the risks from complex mixtures also leads to important choices. EPA's use of a TEQ approach represents the prevailing strategy (in the United States and internationally). In Chapter 3, the committee provides an in-depth evaluation of EPA's use of toxic equivalency factors (TEFs) and TEQs. This issue also represents an important area of uncertainty in the overall risk assessment. The Reassessment states that "despite the uncertainties in the TEF methodology, the use of this methodology decreases the overall uncertainty of the risk assessment" (Reassessment, Part III, p. 1-10). Although that may be true, EPA should quantitatively support the argument with some comparisons or data. The Reassessment also notes that "TEFs are the result of scientific judgment of a panel of experts who used all of the available data, and they are selected to account for uncertainties in the available data and to avoid underestimating risk. In this sense, they can be described as public-health conservative values" (Reassessment, Part III, p. 1-5). The committee recommends that EPA quantify the extent to which the TEF estimation process may be health protective. In addition, because TEFs continue to evolve (see Chapter 3), EPA must continue to choose which TEF values to use and which congeners to include. Such choices will influence exposure estimates as well as the uncertainties associated with those estimates.

The Reassessment acknowledges the difficulty of comparing different human-exposure data sets because some do not include coplanar polychlorinated biphenyls in the estimation of TEQ values. The Reassessment clearly states that TCDD per se is not the main contributor to TEQ levels in human lipids (Part III, Table 4-5). The Reassessment uses the calculation of body burden at steady state, its associated assumptions given in the Reassessment (Part III, section 1.3), best estimates of current adult intakes, and the assumption of 25% body fat to calculate the TEQ concentration in human lipids. The resulting estimate is about one-half the level currently measured in human lipids. The Reassessment suggests that this discrepancy arises from the presence of an historical body burden and lipid concentration, but it does not consider other possibilities.

GENERAL ISSUES RELATED TO VARIABILITY AND UNCERTAINTY ASSOCIATED WITH SELECTION OF DOSE METRIC AND DOSE-RESPONSE MODELING

EPA makes a number of assumptions about the appropriate dose metric and mathematical functions to use in the Reassessment's dose-response analysis (see "Selection of Dose Metric" and "Dose-Response Modeling" in this chapter for specific issues related to dose metric and dose-response modeling). The Reassessment does not adequately comment on the extent to which each of these assumptions could affect the resulting risk estimates.

EPA discussed various dose metrics and selected one particular metric based on its judgment. However, EPA did not quantitatively describe how this particular selection affected its estimates of exposure and therefore provided no overall quantitative perspective on the relative importance of the selection.

EPA faced numerous choices with respect to developing quantitative models for characterizing cancer risk from exposure to TCDD, other dioxins, and DLCs (summarized in Table 2-1) and for characterizing noncancer effects (summarized in Table 2-2). The Reassessment characterizes the risk of cancer at background and incremental intakes by using a cancer slope factor (CSF), and it recommends the use of a margin of exposure (MOE) for both noncancer and cancer end points (Reassessment, Part III, p. 6-12). The committee did not find EPA's justification sufficient for why it used different methods to characterize risk for end points that have the same basic underlying mode of action. The committee noted that the Reassessment should also quantitatively characterize the impact of this choice.

The Reassessment concludes that setting an RfD is not appropriate because of the relatively high background levels compared with effect levels and suggests that setting an RfD provides little value for evaluating possible risk management options if average background exposure exceeds the RfD (Reassessment, Part III, p. 6-14). As discussed in Chapter 7, this decision conflicts with the choices made by other international regulatory bodies (e.g., European Scientific Committee on Food, Food and Agricultural Orga-

TABLE 2-1 Categories of Key Decisions EPA Faced in Characterizing Cancer Risk

Basis for Quantification	Epidemiological Data Set	Bioassay Data Set	Dose-Response Model	Dose Metric	Point of Departure
• Epidemiological and bioassay data • Epidemiological data • Bioassay data • Other	• Choose from individual studies • Use multiple studies	• Choose from individual studies • Use multiple studies	• Low-dose linear • Nonlinear • Multiple • Other	• Average daily dose • Area under the curve • Lifetime average body burden • Peak • Other	• ED_{01} • ED_{05} • ED_{10} • LED_{01} • Other

Abbreviations: ED, effective dose; LED, lower confidence limit on ED.

TABLE 2-2 Categories of Key Decisions EPA Faced in Characterizing Noncancer Risk

Basis for Quantification	Epidemiological Data Set	Bioassay Data Set	POD	Dose Metric	Critical Effect Choice
• Epidemiological and bioassay data • Epidemiological data • Bioassay data • Other	• Choose from individual studies	• Choose from individual studies	• LOAEL • NOAEL • ED_{01} • ED_{05} • ED_{10} • BMD • Other	• BB • ADD • AUC • Peak • Other	• Reproductive and developmental • Immunotoxicity • Neurotoxicity • Central nervous system • Diabetes • Enzymatic change • Other

Additional Categories

Exposure Route	Exposure Time	Type of Dosing	DRD	U.F. (Database)	U.F. (Interspecies)	U.F. (Intraspecies)
• Ingestion • Inhalation • Multiple • Other	• Depends on individual studies	• Single • Multiple	• Yes • No	• 10 • 3 • 1 • Chemical-specific adjustment factor • Other	• 10 • 3 • 1 • Chemical-specific adjustment factor • Other	• 10 • 3 • 1 • Chemical-specific adjustment factor • Other

Abbreviations: POD, point of departure; LOAEL, lowest-observed-adverse-effect level; NOAEL, no-observed-adverse-effect level; ED, effective dose; BMD, benchmark dose; BB, body burden; ADD, average daily dose; AUC, area under curve; DRD, develop reference dose; U.F., uncertainty factor.

nization of the United Nations [FAO]/World Health Organization [WHO], and the Joint Expert Committee on Food Additives [JEFCA]). EPA's decision not to specify an RfD in the Reassessment may have depended on the set of specific assumptions it selected, such as use of the 1% effective dose (ED_{01}) as the point of departure (POD) for this calculation and the magnitude of the applicable uncertainty factors.

The Reassessment provides a thorough statement of the potential sources of uncertainty for consideration in noncancer risk assessment, many of which also apply in the context of cancer risk assessment:

> Consideration should be given to a number of difficulties and uncertainties associated with comparing the same or different endpoints across species, such as differences in sensitivity of endpoints, times of exposure, exposure routes, and species and strains; the use of multiple or single doses; and variability between studies even for the same response. The estimated ED_{01}s may be influenced by experimental design, suggesting caution should be used when comparing values from different designs. Caution should also be used when comparing studies that extrapolate ED_{01}s outside the experimental range. Furthermore, it may be difficult to compare values across endpoints. For example, the human health risk for a 1% change of body weight may not be equivalent to a 1% change in enzyme activity. Similarly, a 1% change in response in a population for a dichotomous endpoint is different from a 1% change in a continuous endpoint, where the upper bound of possible values may be very large, leading to a proportional increase in what constitutes the 1% effect level. Finally, background exposures are often not considered in these calculations simply because they were not known. (Reassessment, Part III, p. 5-24)

The Reassessment used empirical, full dose-response modeling to estimate PODs, specifically an ED for cancer and noncancer. Historically, a POD for a noncancer end point was based on a no-observed-adverse-effect level (NOAEL) or a lowest-observed-adverse-effect-level (LOAEL), a practice inconsistent with cancer risk assessment. EPA now recommends the use of a benchmark dose (BMD) approach to derive a POD for noncancer end points. Although a lower confidence bound on an ED was cited in the literature to define a BMD, EPA's BMD guidance document (EPA 2000b) defines the ED, BMD, and the lower one-sided confidence limit on the BMD (BMDL).[3] This definition unified the determination of PODs for

[3]"BMD is used generically to refer to the benchmark dose approach; in the more specific cases, BMD ... refer[s] to the central estimates, for example the EDx ... for dichotomous endpoints (with x referring to some level of response above background, e.g., 5% or 10%). BMDL ... refers to the corresponding lower limit of a one-sided 95% confidence interval on the BMD...." (EPA 2000b, Executive Summary)

cancer and noncancer end points. The modeling process generally involves two steps:

> The first step is an analysis of dose and response in the range of observation of the experimental or epidemiologic studies. The modeling yields a POD near the lower end of the observed range, without significant extrapolation to lower doses. The second step is extrapolation to lower doses. The extrapolation approach considers what is known about the agent's mode of action. Both linear and nonlinear approaches are available. (EPA 2005a)

This analysis requires making several key decisions, including primarily (1) determining appropriate types of studies (epidemiological, animal, both, and other), (2) choosing specific studies and subsets of data (e.g., species and gender), (3) choosing specific end points for dose-response modeling, (4) choosing a specific dose metric, (5) choosing model type and form, (6) selecting the benchmark response (BMR) and POD, and (7) characterizing uncertainty.

Current EPA practice generally relies on choosing to model a single data set, specifically the one that tends to show the most significant potential adverse effect. This choice can introduce substantial uncertainty into the risk estimation process, particularly in cases in which different data sets yield very different results. One way to avoid the uncertainty introduced by the selection of a single data set is to use multiple data sets. In particular, EPA could place some weight on each of a number of data sets. Chapters 5 and 6 review EPA's data set choices made in the Reassessment.

GENERAL ISSUES RELATED TO RISK CHARACTERIZATION

Critical issues related to risk characterization (see Chapter 7) include the impact of decisions on the information communicated to risk managers about the magnitude of uncertainties associated with the data used to generate risk estimates. The impact of choices made in the risk assessment process can be characterized by quantifying the impact of plausible alternative assumptions at critical steps. The risk estimates can be most fully characterized by performing probabilistic analyses when possible and by presenting the range of possible risk estimates rather than by reporting the single point estimates. Risk characterization should provide useful information to risk managers to help them understand the variability and uncertainty in the risk estimates. As further discussed in Chapter 5, the committee understands that quantitatively addressing all sources of uncertainty in a risk assessment can impose an analytical burden, which may result in addressing some sources of uncertainty qualitatively. Quantifying the contribution of various assumptions to the overall uncertainty often proceeds in an iterative manner. The process itself adds value by highlighting opportunities to collect valuable information, and NRC (1994) provides some guidance about at what point it makes sense to stop in the context of characterizing risks to inform risk management decisions.

The rationale and scientific basis for important decisions should be described in the Reassessment and the consequences of alternative assumptions explored. For dioxin, these issues are best illustrated in relation to the estimation of cancer risk. The choice of one possible approach, linear extrapolation from a POD, results in a CSF that could be used to estimate the lifetime cancer risk for the U.S. population. Assessing the same epidemiological data with a MOE approach would describe the data available to quantify the POD and exposure but would avoid the scientifically debatable need to generate a slope factor with its inherent uncertainties (see Chapter 5 for full discussion of these issues). For noncancer end points, the hazard characterization data are tabulated, but EPA makes little attempt to interpret or focus on critical effects or to define the strengths, weaknesses, and uncertainties associated with effects relevant to critical life stages such as in utero exposure (see Chapter 6 for full discussion of noncancer end points).

The reality that the risk assessment process for TCDD, other dioxins, and DLCs now extends over a period of 14 years, with multiple EPA reports and iterations of these reports, leads the committee to suggest that EPA should continue to treat the risk assessment as a process. In this context, EPA should expect to continue to iterate and improve on the assessment over time as new information becomes available. However, instead of producing and continuing to add to massive reports, EPA should consider a database structure that will allow it to focus its reports on syntheses of new information that drive the quantitative estimates of risk rather than on cataloging all information.

In addition, the committee expects that EPA could substantially improve its assessment process if it more rigorously evaluated the quality of each study in the database. As an example, Table 2-3 summarizes one approach used to describe the basic elements of conducting a systematic review of scientific evidence. Although EPA performed many of these steps in its evaluation of the epidemiological literature of carcinogenicity, it did not outline eligibility requirements or otherwise provide the criteria used to assess the methodological quality of other included studies. EPA could also substantially improve the clarity and presentation of the risk assessment process for TCDD, other dioxins, and DLCs by using a summary table or a simple summary graphical representation of the key data sets and assumptions (e.g., using trees like those shown by Evans et al. 1994a,b; Sangrujee et al. 2003).

TABLE 2-3 Components of a Systematic Review

- State objectives and hypotheses
- Outline eligibility criteria, stating types of study, types of participants, types of interventions and outcomes to be examined
- Perform a comprehensive search for potentially eligible studies
- Decide eligibility and assess methodological quality of included studies
- Tabulate study characteristics
- Extract data, with involvement of investigators if necessary
- Analyse results of included studies, using statistical synthesis of data (meta-analysis), if appropriate
- Prepare a report of review, stating aims, materials and methods and describing results and conclusions

SOURCE: Smyth 2000.

SELECTION OF DOSE METRIC

Section 1.3 of the Reassessment Part III considers various dose metrics for understanding exposure and analyzing dose response relationships, which apply to both cancer and noncancer effects. EPA highlights the need for a pragmatic approach that can be applied to issues of cross-species scaling and to different end points detected under different exposure scenarios. Risk assessments for most chemicals typically focus on the external dose or exposure expressed as mass of substance per kilogram of body weight per day, but many other options exist. The Reassessment discusses a number of different dose metrics that represent the internal dose, including estimates of area under the blood or plasma concentration–time curve (AUC), plasma or tissue concentrations, body burden, and function-related biomarkers of the internal dose such as aromatic hydrocarbon receptor (AHR) occupancy or changes in cytochromes P450A1/2 protein (CYP1A1/2) activity. The function-related biomarkers are intellectually appealing, especially for extrapolating from animal to human, because they would provide a means to address species differences in toxicokinetics and in the initial events reflecting tissue sensitivity. However, EPA concluded that insufficient data support the current use of function-related biomarkers in risk assessment.

The Reassessment (Part III, p. 1-17) suggests that, at the present time, body burden represents the most suitable dose metric for interspecies comparisons (similar to the approaches used by other recent evaluations of TCDD, other dioxins, and DLCs [SCF 2000, 2001; JECFA 2002]), while lifetime AUC may also be suitable for comparisons of different human exposures. EPA selected body burden for cross-species comparisons because, "assuming similar sensitivity between rats and humans at the tissue level, effective doses should be a function of tissue concentration," and "tissue concentrations of TCDD and related chemicals are directly related to the concentration of TCDD in the body" (Part III, p. 1-12).

Chapter 5 discusses the quantitative importance of this assumption in terms of cancer risk assessment and provides additional discussion of alternative dose metrics and the relative importance of the choice of dose metric on ultimate cancer risk projections.

The Reassessment states, "The steady-state concentration of TCDD in the body, or steady-state body burden, can be estimated in rats and humans using the following equation:

$$\text{Steady-state body burden} = \frac{\left[\text{Dose (ng TEQ/kg)} \times t_{\frac{1}{2}}\ \text{(days)}\right] \times F}{Ln(2)}, \qquad 2\text{-}1$$

where Dose is the daily administered dose, F is the fraction absorbed, and $t_{1/2}$ is the species-specific half-life of TCDD" (Reassessment, Part III, p. 1-12). Body burdens after shorter periods of administration (non-steady state) would require a different method of estimation.

The Reassessment does not quantitatively explore the impacts of this choice or the choices of various inputs in the equation (see below) used to estimate body burden at steady state. The summary table in the Reassessment (Part II, Table 1-6) gives limited data for the half-life estimates for TCDD. Estimates of elimination half-lives for various tissues in rats range from 11 to 53 days, with the best data coming from eight studies that used a radiolabeled compound and that reported a range of 12 to 31 days. EPA uses 25 days to calculate the body burden in rats at steady state. That appears appropriate to the committee, but this estimate is clearly uncertain. Similarly, the summary table in the Reassessment (Part II, Table 1-10) gives limited data on half-life for TCDD in humans. The table provides an estimate of 5.8 years based on fecal excretion and 9.7 years based on changes in adipose concentrations. Data from the Operation Ranch Hand Study indicated TCDD half-lives of 7.1 (Michalek et al. 1992) and 11.3 years (Wolfe et al. 1994), the most comprehensive recent analyses indicating a half-life of 7.6 years (95% confidence interval of 7.0 to 8.2 years) (Michalek and Tripathi 1999). The Reassessment (Part II, Table 1-13) reports a half-life of 7.2 years for the Flesch-Janys et al. (1996) study. An overall mean serum TCDD half-life of 8.2 years was reported in 27 victims of the accident in Seveso, Italy (Needham et al. 1994), although a recent study found substantial interindividual variability and concentration-dependent differences in TCDD half-life (Aylward et al. 2005). Overall the value of 2,593 days (or 7.1 years) used by EPA to calculate the body burden in adult humans at steady state appears reasonable and realistic. The Reassessment recognizes that TCDD half-life is shorter in neonates and infants. The

Reassessment notes that TCDD half-life varies with percent body fat and increases significantly with a high percent of body fat, suggesting that people with more body fat tend to eliminate TCDD more slowly. The half-life of TCDD shows a significant correlation with body weight (IOM 2000). These two pieces of data indicate that human variability in elimination is related to differences in the apparent volume of distribution as well as clearance (see below). The values for bioavailability used in the above equation are also somewhat uncertain. The summary table in the Reassessment (Part II, Table 1-1) gives only limited data for TCDD in rats showing a high bioavailability (70% and 84% in two studies using acetone, corn oil gavage). The text describes the absorption of 88% of TCDD in male Fischer 344 rats after oral exposure in Emulphor/95% ethanol/water (1:1:3). EPA assumed 50% absorption from the diet for rats, which appears reasonable because a range of 50% to 60% absorbed has been reported. The summary table in the Reassessment (Part II, Table 1-1) gives data from only one study for TCDD in a human given a single oral dose and gives a bioavailability of 87% (Poiger and Schlatter 1986). Other studies have determined the extent of absorption by mass balance (the amount ingested minus the amount eliminated in feces), but such measurements are likely to be unreliable in adults because elimination of unchanged TCDD in feces is an important route of elimination of absorbed TCDD in humans. Overall the value proposed and used by EPA to calculate the body burden in humans at steady state (80% absorption) appears reasonable, although the data are limited.

Equation 2-1 implicitly assumes that body burden represents a good surrogate for tissue concentration and that adverse effects correlate with steady-state body burden. This assumption represents a reasonable default because the body burden generally appears to be proportional to tissue concentration, with some caveats noted in Chapter 5, and the toxic effects of TCDD, other dioxins, and DLCs increase with increased tissue concentration. However, the use of body burden as a dose metric (or a dose metric based on tissue concentration) would not allow for species differences in inherent target organ sensitivity to the presence of the chemical. Species differences in target organ sensitivity could be taken into account by a full biologically based kinetic-dynamic model, but EPA appropriately concluded that the available models remain insufficiently well validated for risk assessment purposes. The committee did not discuss specific recommendations for EPA related to collecting data for refining current BBDR models or the regional induction models, but the committee encourages further development and use of these models as data become available to validate and further develop them.

The use of Equation 2-1 implies comparable steady-state tissue concentrations between species and between individuals simply on the basis of body burden. Assuming dose linearity, a twofold increase in body burden in any individual will yield a twofold increase in the concentrations in all tissues, but the actual concentrations in any tissue will depend on the pattern and extent of tissue distribution of the total body burden.

Equation 2-1 implies that different half-life values between and within species will result in different body burdens for the same daily intake. However, the Reassessment does not explicitly characterize how different half-life value choices influence risk estimates. The half-life depends on two independent physiological variables: the clearance (CL), which reflects the volume of blood cleared per unit time, and the apparent volume of distribution (V), which reflects the apparent volume of blood that has to be cleared of chemical and which is determined by the extent of distribution to tissues (for the one-compartment model used by EPA, half-life = 0.693 × V/CL). The half-life, and therefore the estimated body burden at steady state, could differ between species or between individuals due to differences in clearance or in the extent of tissue distribution (V)—for example, due to differences in body fat content. Because half-life depends on both CL and V, and body fat content represents the major determinant of V for TCDD and other dioxins, a species with a proportionately higher body fat content would have a proportionately higher value of V, a proportionately longer half-life, and greater body burden at steady state for the same daily intake. For this reason, the blood concentration at steady state offers a better metric of the concentration available within tissues to produce an effect:

$$\text{Steady-state concentration} = \frac{\text{Daily Dose} \times \text{Bioavailability}}{\text{CL}}, \qquad 2\text{-}2$$

where concentration means the concentration per unit volume in blood or plasma, and CL is expressed as the volume of blood or plasma cleared of chemical per day.

This equation cannot be readily used because no data are available on CL for humans. (CL would be the sum of all processes that remove the compound from the body, which in the case of TCDD would largely relate to diffusion into fecal lipids, whereas for lower chlorinated congeners, the value of CL would also reflect metabolism.)

The Reassessment (Part III, section 1.3.2) considers the possibility of using AUC as a dose metric, especially for the purpose of estimating cancer risk. However, EPA questions the use of AUC because animal studies show more altered hepatic foci after a single high dose than after repeated low-dose exposures giving the same AUC and because of challenges in determining the appropriate averaging time (e.g., the whole lifetime or some discrete window of susceptibility). The Reassessment notes that species life-span differences imply a time-based correction to AUC across species, the correction making AUC equivalent to average steady-state concentration. EPA could convert the AUC over any period to an average concentration by dividing by the time period. The AUC for a dose interval at steady state is directly proportional to the daily dose and bioavailability divided by CL because $AUC_{\text{dose interval at steady state}}$ = (dose × bioavailability)/CL. The apparent volume of distribution does not influence blood or plasma AUC for a dose interval at steady state, unlike body burden. The blood or plasma concentrations would not vary greatly during a dose interval (day) because of the long half-life of TCDD in both rodents and humans, and therefore the average blood or plasma concentration could be used. The criticism of using AUC in the Reassessment (whether it should be the peak AUC or the average AUC related to the toxic effect) is inappropriate because it applies equally well to the body burden metric used by EPA.

The Reassessment (Part III, section 1.3.3) considers the use of plasma or tissue concentrations as a dose metric and states that few such data exist for the chronic and subchronic animal studies, whereas human exposure data depend predominantly on such measurements. The human data expressed on a lipid-adjusted basis complicate interspecies comparisons with rodent plasma data, and few data are available to quantify tissue concentrations during toxicity studies in animals. If possible, direct comparisons of the concentrations in the lipid fraction of human blood and rodent blood would provide the most secure comparison of internal dose if such data became available in the future. Tissue concentration data for animals and humans could be developed with physiologically based pharmacokinetic (PBPK) models, based on the proportion of body fat and data on organ blood flows and partition coefficients. Differentiation of free compound from lipid-bound compound within a PBPK model could provide the most relevant dose metric for dose-response assessment.

The approximately 100-fold difference between rats and humans in TCDD half-life combined with Equation 2-1 suggests that a 100-fold lower daily intake in humans yields a total body burden equal to that in rats (assuming the same bioavailability). This observation raises a key question not considered adequately in the Reassessment: Would similar *total* body burdens in rats and humans result in similar target organ concentrations? Similar tissue concentrations in both species would occur if the pattern of distribution of the body burden were the same in both species. However, the extent of hepatic sequestration (higher in rats, see Reassessment, Part II, Tables 1-4 and 1-5) and the proportion of body fat (10% of body weight in rats according to Geyer et al. [1990] and about 25% in humans—see Reassessment, Part III, p. 17) both show important differences between rats and humans. The significance of the different body composition can be illustrated by considering the TCDD concentrations in rats and humans that would be associated with a total body burden of 200 ng/kg of body weight (calculated from the intake and half-life), assuming a body fat/blood concentration ratio of 100:1 at equilibrium for both rats and humans, and that body fat is 10% of body mass in rats and 25% of body mass in humans. For rats, the fraction of the body burden of TCDD in fat would be proportional to 100 × 0.1 (10), and the fraction of the body burden of TCDD in nonfat tissues would be proportional to 1 × 0.9 (0.9). Hence, a total of 183.5 ng of TCDD would be in fat, and 16.5 ng would be in nonfat. The total concentrations are 1,835 ng/kg in fat and 18.3 ng/kg in nonfat tissues. In humans, TCDD in fat would be proportional to 100 × 0.25 (25), and the amount of TCDD in nonfat tissue would be proportional to 1 × 0.75 (0.75). Therefore, for a body burden of 200 ng/kg of body weight, the total TCDD in fat would be 194.2 ng, giving a TCDD concentration of 776.7 ng/kg, and the total in nonfat tissue would be 5.8 ng, giving a concentration of 7.8 ng/kg. Consequently, for the same total body burden, the TCDD and other dioxins concentrations in the tissues of humans are about two to three times lower than those in rats.

The higher hepatic uptake in rats compared with humans means that, for the same total body burden, there would be a greater proportion of TCDD in the livers of rats. The Reassessment applies the same body burden correction factor between rats and humans for liver cancer and for nonhepatic effects. The proportionately higher concentrations in the livers of rats compared with humans means that a proportionately higher daily intake would be necessary in humans to produce a comparable hepatic concentration. The difference in hepatic concentration based on the use of body burden as a dose metric for extrapolation of data on liver cancer in rodent bioassays to humans would represent an assumption that makes the resulting risk estimate conservative, although the implications of this assumption are not described in the Reassessment. In addition, the Reassessment does not consider alternative assumptions. Because of the difference in the percent of body fat, the same overall TCDD body burden generally corresponds to lower tissue concentrations in humans, a factor that makes extrapolation of data for all effects (including hepatic effects) more conservative. The Reassessment does not address this factor.

The tissue distribution of the body burden in studies that used single doses or short periods of treatment will not correspond to the steady-state pattern. Before completion of the distribution phase, there will be higher concentrations in well-perfused tissues and lower concentrations in adipose tissue. JECFA (2002) allowed for such nonequilibrium distribution in its recent evaluation of the in utero effects produced in rats shortly after a single dose of TCDD. The EPA Reassessment did not consider this approach in the body burden calculations for the same studies.

The Reassessment does not adequately consider the use of a PBPK model to define species differences in tissue distribution in relation to total body burden for either cancer or noncancer end points. Kim et al. (2002)

compared the body burdens associated with different levels of biochemical responses calculated using a simple kinetic approach and using the body burden derived from a PBPK model. The results indicated that the simple kinetic method, which was similar to that used by EPA, and the PBPK model gave quantitatively different results. The differences were not consistent across the biochemical end points studied, suggesting that the response model used was influencing the magnitude of the difference. Nevertheless, this study supports the conclusion by the committee that the Reassessment should use a simple PBPK model to address some of the uncertainties inherent in the use of species differences in body burden as a measure of species differences in target organ exposure. Generic PBPK models and PBPK models developed specifically for TCDD and its congeners incorporate about 7% of the body weight present as adipose tissue in rats and about 15% in humans (Gerlowski and Jain 1983; Wang et al. 1997; Maruyama et al. 2002, 2003; Emond et al. 2004). Simple PBPK models of TCDD biodisposition at steady state could be used to convert the estimated body burden into an appropriate species-related difference in steady-state tissue concentrations; the magnitude of the resulting species difference could then be introduced as a correction factor in the equation used by EPA to calculate body burden from intake, half life, and bioavailability. The same PBPK model might also be used to explore the influence of human variability in body composition on the elimination half-life and therefore the body burden at steady state. The Reassessment did not consider this approach or quantify its impact, despite its recognition of tissue concentration as the best dose metric.

DOSE-RESPONSE MODELING

Background

A critical element to consider when assessing human variability in response to a toxic substance is the nature of the dose-response relationship, and how it is modeled mathematically. As described in major textbooks in toxicology (e.g., Eaton and Klaassen 2001), analysts model two fundamental types of dose-response relationships. The graded (continuous), individual dose response characterizes the nature and magnitude of an individual's response to a toxic substance as the dose goes from a small, ineffectual dose to a larger, toxic dose, potentially causing death. The nature of the response may differ qualitatively, depending on the dose and duration of exposure. For any given individual and specific, defined effect, a "threshold dose," may exist, which is defined as the dose below which the individual does not respond. The dose corresponding to that threshold may differ across individuals. For the purposes of risk assessment and public health protection, however, analysts typically use the second type of dose-response relationship, called the "quantal dose-response relationship," for a population of exposed individuals. The quantal dose response describes the relationship between exposure and the proportion of the population that will exhibit a health effect (that is, a separate relationship for each adverse end point).

In the case of TCDD, other dioxins, and DLCs, it is important to assess the population-based dose-response relationship for cancer, birth defects, immunotoxic effects, and so forth. For each end point of interest, individuals in a population (e.g., rats and mice in laboratory studies and humans in epidemiological studies) are identified as either responders or nonresponders at defined doses (quantal responses). The cumulative quantal dose-response relationship for the population is then determined from the distribution of responses in the population across a defined range of doses. The term threshold is often used to describe the dose below which no response occurs for the graded (continuous) dose-response relationship or the dose below which the probability of anyone in the population responding approaches zero for the cumulative quantal dose-response relationship. A common but scientifically unachievable goal in risk assessment is to identify a threshold dose that protects everyone in the population. The term offers some value in recognizing that for the vast majority of dose-response relationships (either individual or population) some doses may exist below which no measurable responses occur (in an individual or a population). However, the term threshold remains subject to many vagaries of interpretation, and the committee prefers to express ranges of dose in terms of MOEs. MOEs are usually defined as the ratio of the highest dose (daily exposure) to an agent presumably without adverse impact on the human population (the so-called reference dose; Faustman and Omenn 2001) to the estimated daily human dose that might occur, determined from analysis of actual exposure scenarios.

Because of inherent biological differences between individuals, as well as the probabilistic nature of many toxic responses, distributions in responses in a population will always exist (that is, not everyone responds the same way to the same dose). In human populations, differences arise from genetic diversity, differences in age, gender, nutritional status, diseases, and other concomitant exposures, which can modify the response of an individual to a toxic substance. However, such contributors to human variability are presumably represented in the data sets obtained in human population-based studies (epidemiological studies), although any one study generally cannot capture the full range of possible individual variability in response. A second major challenge in establishing population-based dose-response relationships in epidemiological studies arises from the frequently poor quality of exposure (dose) information. Although well-designed occupational and environmental epidemiological studies can yield useful information on human population variability, relatively little quantitative information is available about the potential impact on genetic polymorphisms in the human population that might give rise to differences in susceptibility to the toxic effects of TCDD, other dioxins, and DLCs. Chapters 5 and 7 provide more discussion about genetic, molecular, and biochemical mechanisms that might contribute to interindividual variation in response to TCDD, other dioxins, and DLCs.

With these caveats noted, risk assessors commonly take existing data sets (both animal and human) and attempt to develop mathematical models to characterize the shape of the dose-response relationships from the observed data.

Dose-response modeling is a process to formally quantify dose-related changes in the incidence or severity of an adverse effect. The scale of the response can be quantal (e.g., cancer incidence) or continuous (e.g., AHR-binding immune response). Analysts use mathematical functions (preferably with mechanistic parameters) to describe the dose-response relationship observed in the data. In the case of cancer or any quantal outcome, the dose-response model, R(dose), is the same as the probabilistic risk of the adverse outcome. With this dose-response model, or risk, R(dose), the ED_{α}, at which there is a prespecified, small amount (typically 1 ~ 10%) of risk increase α above the background, can be estimated by the following equation of excess risk:

$$\frac{R(ED_{\alpha}) - R(\text{background exposure})}{1 - R(\text{background exposure})} = \alpha.$$

The risk increase α is called the effective dose level. Because R(dose) is a statistically estimated quantity (function), the resultant ED_{α} is subject to data variation.

In the case of a continuous response (or more generally, a nonquantal response), EPA guidance documents discuss how the type of data and biological knowledge will determine appropriate methods using general approaches, but no single approach or model can be universally the "best." Analysts first fit a dose-response model R(d) to the response data. They then take additional steps to formulate a measure of risk based on the model. Here, R(d) describes the mean response level of the toxicological outcome (e.g., cognitive function as measured in terms of IQ test score in the case of exposure to a neurotoxin). The Reassessment discusses several proposed approaches (Part II, pp. 15-16), all of which identify a dose associated with a specified level of response change relative to the control. For continuous responses, this task is complicated by the ambiguous separation between a "normal response" and an "adverse response." In lieu of an obvious dividing line, EPA used the "dynamic range" approach (Murrell et al. 1998), which defines ED_α (EPA assigned an α value of 1%) as the dose satisfying the relationship,

$$\frac{R(ED_\alpha) - R(background)}{R_{max}} = \alpha.$$

where R_{max} is the maximum range of total response, either theoretical or estimated under the maximum exposure condition. The main drawback of the approach used by EPA is that the response level associated with the ED_α may not be clinically or toxicologically important. The NRC (2000) described an alternative approach in the context of its review of methylmercury toxicity, based on work by Crump (1984) and Gaylor and Slikker (1992). That approach first identifies an adverse response level, which demarcates normal and abnormal (or adverse) responses. For example, in the case of a neurotoxin, an IQ score of 70 points (two standard deviations [SDs] below the population mean of 100 points) could be designated the adverse response level because individuals with IQ scores below this level often require community support to live (WHO 1992, as cited in EPA 2005c). The ED_α is then defined to be the neurotoxin dose that increases the background probability of an adverse response by α. Continuing the IQ example, the ED_{05} is the level of neurotoxin exposure that increases the background risk of having an IQ below 70 of 2.5% by an extra 5% (5%*97.5%=4.875%), to a total of 7.375%.

The Reassessment (Part II, p. 8-16) identifies difficulties with this approach. Although such an adverse response level might not always identify toxicologically meaningful events, it can identify unusual outcomes outside the normal range. The committee recognizes this challenge and understands that for some end points this may emerge as an insurmountable challenge. Nonetheless, because the ED_α definition used by EPA is difficult to interpret toxicologically, EPA should strive to use the alternative approach described here whenever possible.

Historically, risk assessment of noncancer effects used a NOAEL or a LOAEL as the POD. The BMD approach (Crump 1984) eliminates some of the limitations of the NOAEL and LOAEL approach and makes the analysis of noncancer effects more consistent with that of cancer.

The primary objective of dose-response modeling is to define an ED toward the lower end of the experimental dose range where the model remains supported by adequate data. The ED can then be used as a POD for extrapolation toward an environmental background level or for safety assessment using the MOE approach.

The choice of model for dose-response assessment, choice of the POD, and extrapolation below the POD thus represent other key areas of uncertainty. The Reassessment quantified the cancer dose-response relationship relying primarily on occupational cohort data. EPA also used selected animal bioassay data to confirm the plausibility of the resulting estimates. Specific issues related to choice of data set for cancer risk assessment are discussed in more detail in Chapter 5

Data Set Selection

Full dose-response modeling requires adequate dose-response data, and adequate selection criteria must be applied. EPA's guidance document (EPA, 2000b, p. 14) states:

> In general, studies with more dose groups and a graded monotonic response with dose will be more useful for BMD analysis.... Studies in which responses are only at the same level as background or at or near the maximal response level are not considered adequate for BMD analysis. It is preferable to have studies with one or more doses near the level of the BMR to give a better estimate of the BMD and, thus, a shorter confidence interval. Studies in which all dose levels show changes compared with control values (i.e., no NOAEL) are readily useable in BMD analyses, unless the lowest response level is much higher than that at the BMR.

Depending on whether the scale of the selected end point is quantal (dichotomous), continuous, or categorical, different statistical procedures and models are required for dose-response modeling.

EPA's Reassessment selected a large body of published data sets, using the criteria of (1) a positive dose trend and (2) at least three dose groups in addition to a control (more specifically for noncancer data). In dose-response modeling of human cancer data, EPA further used cancer death incidence (time-to-event) data as the end point, which generally provides more information than mortality data by considering when a death occurred. (These studies are discussed in more detail in Chapter 5.)

Statistical Power and Precision

Although meeting those minimal selection criteria (discussed above) is critical, it does not guarantee adequate statistical power to ascertain the shape of the dose-response curve, and it does not account for the associated uncertainty. In the present context, statistical power refers to the general ability of an experiment, and its associated data set, to provide information needed to make a reliable inference, including testing positive dose effects and ascertaining a fitted dose-response model.

The Reassessment did not discuss the issue of statistical power, although the cancer guidelines (EPA 2005a, see also Appendix B) recommend assessing the statistical power of the studies used for dose-response assessment when possible. Even if a study possesses adequate statistical power to confirm a positive overall dose response within the observed data range, the power might be inadequate to ascertain the shape of the dose-response curve below the POD level. The lack of statistical power at the lower end also represents a problem for both cancer and many of the noncancer data sets, contributing additional uncertainty to the POD.

Choice of the Dose-Response Model

The goal of mathematical modeling in determining a POD is to fit a model that describes the data set well, especially at the lower end of the observable dose-response range. Fitting such a model involves first selecting models for consideration, based on the characteristics of the data and experimental design, and then fitting the models using one of a few established methods. Then, an ED, along with its upper and lower confidence bounds, is calculated at the POD level. In the process, the analysis should evaluate model fitting, determine goodness-of-fit, and compare models to decide which one to use for obtaining the POD. For example, the BMD guidance document (EPA 2000b) recommends use of a *P* value of 0.1 as the reference critical value for goodness-of-fit (instead of the more conventional values of 0.05 and 0.01), examination of a graphical display of the model fit, and use of Akaike's information criterion for comparison of models and selection of the model to use.

In the case of human cancer data, the Reassessment included fits of linear and nonlinear models to the data (see Chapter 5). With the rodent cancer data, EPA used a simple multistage model fitted with the BMD software program. For noncancer data, EPA used the Hill model as the default for continuous responses, with a power model as the alternative when the Hill model failed to fit the data computationally. (See Chapter 6 for additional discussion about specific noncancer end-point modeling.) EPA used the Weibull model as the default for quantal noncancer data. The committee commends EPA for using flexible mathematical models (e.g., the Hill and Weibull models) to account for both nonlinear and linear shapes of the dose response for noncancer effects. However, the committee recommends that EPA apply similar efforts in dose-response modeling of human cancer data (see Chapter 5).

The Reassessment did not conduct or report statistical tests of goodness-of-fit of the cancer risk models. Two reasons might explain the absence of these test results. First, EPA relied on the models reported in the original publications. For example, Steenland et al. (2001) fitted several models to the risk ratio for cancer death incidence, including a power and a piecewise linear model. The likelihood ratio test showed a statistically significant, positive dose response, but the graphical display clearly showed a potential lack of fit. It is important to note that a higher statistical signifi-

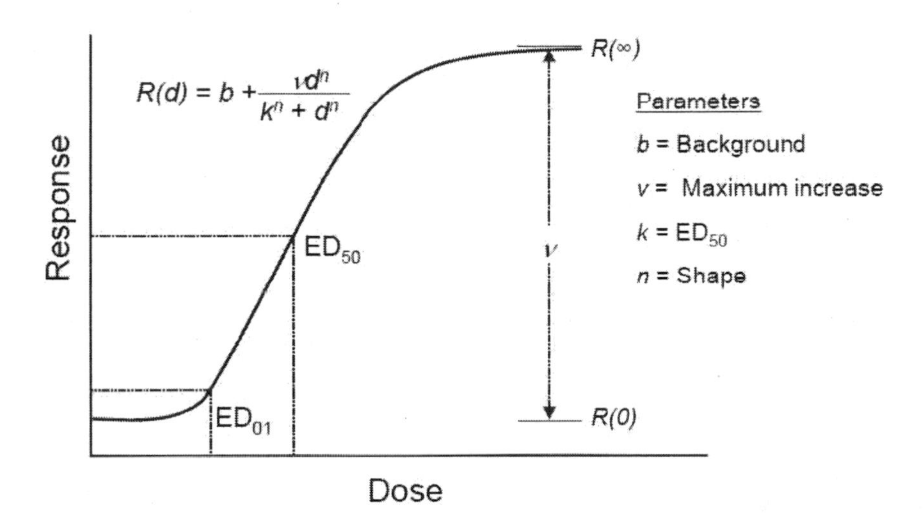

FIGURE 2-1 Vmax. As used in the BMD software for modeling dose-response data, the term Vmax refers to the modeled maximum percent response seen in the observed data set. SOURCE: N. Walker, NIEHS.

cance does not correspond to a higher degree of goodness-of-fit of the model to the data. The Reassessment did not distinguish statistical tests of significance from tests of goodness-of-fit. Second, EPA had access only to summary data taken from the published literature for dose-response modeling, not the raw data, and consequently may not have been able to conduct statistical tests for goodness-of-fit. Nonetheless, the committee recognizes that the critical choice of the dose-response model would benefit from as much information as possible.

In contrast, EPA adopted an ad hoc method to assess goodness-of-fit in dose-response modeling of noncancer end points. Specifically,

> the model fits were evaluated with regard to the observed data. The goodness of the model fit was determined as 'good' if the model curve included nearly all of the data point means, 'marginal' if the model curve was within one standard deviation of the data point means, or 'poor' if model fit was not within one standard deviation of the means.

Furthermore,

> for the Hill model fits, the Vmax [see Figure 2-1] estimates from 'good' and 'marginal' model fits were subjectively evaluated for stability and biological plausibility with regard to the observed data. This evaluation identified some potential problems with some of the Vmax estimates. In

> some cases the error associated with the Vmax could not be calculated by the BMD software. In these cases if the Vmax model estimate was similar to the 'observed Vmax' (i.e. the difference between the highest dose response level and the control response level) then the Vmax estimate was considered biologically plausible and was used for the calculation of an ED_{01}. Otherwise the 'observed Vmax' was used for calculation of the ED_{01}. (Part II, p. 8-32)

This subjective approach to goodness-of-fit did not identify whether the lack of fit occurs at the higher or lower end of the observed dose-response range. Alternatively, the Reassessment could judge goodness-of-fit of an empirical dose-response model on mechanistic grounds.

Finally, a statistically well-fit model alone does not guarantee that the model approximates the true but unknown shape of the dose response, especially below the observed dose-response range. With limited data (e.g., about three dose groups for noncancer data) and limited statistical power, many of the data sets (including epidemiological studies) analyzed in the Reassessment do not provide sufficient information to confirm the true shape of the dose-response curve at the ED_{01} level. The committee emphasizes that this critical uncertainty about low-dose extrapolation remains one of the most significant uncertainties; at the same time, it represents an uncertainty that EPA probably will not resolve in the short term. When feasible, mechanistic and statistical information should be used to ascertain the shape of the dose-response curve at lower doses. Minimally, EPA should use rigorous statistical methods to assess model fitting to control and reduce the uncertainty of the POD caused by a poorly fitted model.

Choice of the POD Value

Selection of the ED (BMR) level is critically important in the calculation of an ED (BMD), and therefore, in the determination of a POD or calculation of a MOE. The current cancer guidelines (EPA 2005a, see also Appendix B) and the draft BMD guidance document (EPA 2000b) give detailed recommendations. For quantal data, an excess risk of 10% was chosen as the default level because 10% response is at or near the limit of sensitivity in most cancer bioassays and in some noncancer studies as well. If a study offers greater than usual sensitivity, then a lower level (e.g., 1%) can be used. EPA recommends the 1% BMR level for epidemiological studies primarily because the 1% level is typically within the observed range. In any case, according to the guidance document, the ED_{10} should be reported along with any other possible POD options. EPA's BMD guidance document further recommends:

> For continuous data, if there is an accepted level of change in the endpoint that is considered to be biologically significant then that amount of change is the BMR. Otherwise, if individual data are available and a decision can be made about what individual levels should be considered adverse, the data can be 'dichotomized' based on that cutoff value, and the BMR set as above for quantal data. Alternatively, in the absence of any other idea of what level of response to consider adverse, a change in the mean equal to one control standard deviation (SD) from the control mean can be used. The control SD can be computed including historical control data, but the control mean must be from data concurrent with the treatments being considered. Regardless of which method of defining the BMR is used for a continuous dataset, the effective dose corresponding to one control SD from the control mean response, as would be calculated for the latter definition, should always be presented for comparison purposes. (EPA 2000b, p. vii)

In EPA's computation of ED_{01} for noncancer continuous end points, the 1% BMR level is defined as the change of response from the background level of the control group that was 1% of the maximum possible total response range. The choice of a 1% BMR level ignored EPA's own guidance that "if there is an accepted level of change in the end point that is considered to be biologically significant then that amount of change is the BMR" (EPA 2000b, vii). The Reassessment also did not consider an alternative approach to dichotomize a continuous outcome into normal and extreme outcomes below a lower or above an upper percentile (Gaylor and Slikker 1990), an approach recommended in the BMD guidance document (EPA 2000b) and implemented in EPA's BMD software program.

Because the shape of the dose-response is less certain at the lower end of the experimental range, the consequent uncertainty for the ED chosen in this range is important. This uncertainty is likely to be greater for the lower confidence bound of ED_{01} than on the central estimate of ED_{01} itself. The Reassessment appears to have largely ignored this issue.

As the starting point of extrapolation of risk to environmental exposure levels, the POD directly influences the risk estimate. The lack of fit of the model at the lower end of the dose-response curve leads to substantial extrapolation of the model toward the POD, and that can bias the ED or BMD estimates and widen their confidence intervals, adding substantially to the uncertainty of the estimate.[4]

---

[4]The accuracy of any experimental measurement is limited by the ability to measure the phenomenon, by any methodological errors introduced through sampling (e.g., limitations in sample size or selection), and by assumptions made in fitting a model to the data. As such, any result obtained provides an estimate of the "true value" with some associated uncertainty. A confidence interval represents the likelihood that the "true value" will occur within

Despite the Reassessment's consideration of multiple options and the use of flexible model forms (such as the Hill model and the Weibull model) to test for nonlinear dose response, mechanistic knowledge gaps, data gaps, and model gaps remain. For example, many of the data sets of noncancer effects yielded a Hill coefficient greater than 1.5, indicating a plausible nonlinear dose response. However, those studies lacked adequate statistical power to estimate the Hill coefficient reliably, rendering the estimate statistically nonsignificant (that is, the confidence interval includes unity). This result represents a general data gap because the dose-response data required to establish a nonlinear dose-response form do not exist, a problem that becomes magnified in extending nonlinear models to the low-dose range. At present, mechanistic knowledge of both cancer and noncancer effects supports the plausibility of a nonlinear dose response at the lower range (see also Chapter 5), but no adequate data or widely accepted dose-response models describe the shape below a chosen POD at or below the 1% level. It is useful to differentiate the lack of data to confirm the shape of the dose-response curve below the POD from the lack of qualitative evidence of nonlinearity. On the whole, the committee concluded that the empirical evidence supports a nonlinear dose response below the ED_{01}, while acknowledging that the possibility of a linear response cannot be completely ruled out. The Reassessment emphasizes the lack of such nonlinear models, hence its adoption of the approach of linear extrapolation below the POD level. Although this approach remains consistent with the cancer guidelines (EPA 2005a, see also Appendix B), EPA should acknowledge the qualitative evidence of a nonlinear dose response in a more balanced way, continue to fill in the quantitative data gaps, and look for opportunities to incorporate mechanistic information as it becomes available. The committee recommends adopting both linear and nonlinear methods of risk characterization to account for the uncertainty of dose-response relationship shape below ED_{01}.

With respect to dose-response modeling, the committee recommends that the Reassessment explicitly acknowledge the lack of statistical power (precision) of the data to estimate the ED_{01} or test nonlinearity of the dose response below the POD level of choice (e.g., ED_{01}).

The committee notes that the choice of the 1% response level as the POD substantially affects both the cancer and the noncancer analyses,

the range of the lower and upper confidence bound. For example, statisticians often choose to report a 95% confidence interval, which implies a 95% chance that the true value will fall within the stated range, but this represents a subjective choice and other choices (e.g., 90% confidence interval) are equally valid. The confidence interval depends on the underlying variability of the quantity being measured or modeled and the number of samples collected and/or available to fit the data. For any given result, collecting more samples tends to narrow the confidence interval.

perhaps driving EPA's decision not to develop an RfD. The committee recommends that the Reassessment use levels of change that represent clinical adverse effects to define the BMR level for noncancer continuous end points as the basis for an appropriate POD in the assessment of noncancer effects. The Reassessment should also explicitly address the importance of statistical assessment of model fit at the lower end and the difficulties in such assessments, particularly when using summary data from the literature instead of the raw data, although estimates of the impacts of different choices of models would provide valuable information about the role of this uncertainty in driving the risk estimates.

CONCLUSIONS AND RECOMENDATIONS

- Although EPA qualitatively addressed many sources of uncertainty and variability, the Reassessment does not adequately address uncertainty and variability that result from the numerous decisions EPA made in deriving point estimates of cancer risk in the comprehensive risk assessment. In contrast, EPA used concerns about uncertainties and uncertainty factors as part of the justification for not setting an RfD for noncancer effects (see Chapter 7 for further discussion).
- The Reassessment does not provide details about the magnitudes of the various uncertainties surrounding the decisions EPA makes in relation to dose metrics (e.g., the impact of species differences in percentage of body fat on the steady-state concentrations present in nonadipose tissues). The committee recommends that EPA use simple PBPK models to define the magnitude of any differences between humans and rodents in the relationship between total body burden at steady-state concentrations (as calculated from the intake, half-life, bioavailability) and tissue concentrations. The same model could be used to explore human variability in kinetics in relation to elimination half-life. EPA should modify the estimated human equivalent intakes when necessary. Many opportunities exist to further characterize sources of uncertainty and variability related to the dose metric choices, and the committee recommends that EPA provide a clear evaluation of the impacts of possible choices on the risk estimates.
- The committee recommends that EPA make greater use of mechanistic information to assess the biological plausibility of different mathematical models, use more rigorous criteria (e.g., goodness-of-fit tests) and follow its own guidance (EPA 2000b) in deriving a POD, and clearly identify the BMR level of toxicological significance for noncancer end points. Many opportunities exist to further characterize sources of uncertainty and variability related to the POD and extrapolation choices, and the committee recommends that EPA provide a clear evaluation of the impacts of possible choices on the risk estimates.

- The committee notes that EPA would substantially improve its transparency and management of the complexity of the risk assessment of TCDD, other dioxins, and DLCs by creating an ongoing process for clearly identifying and updating the key assumptions that support the quantitative risk assessment. This process would essentially require viewing the risk assessment as an ongoing and iterative effort in which EPA continues to create incentives to obtain and use better information when possible and appropriate.

3

Toxic Equivalency Factors

Assumptions, variability, and uncertainty are well delineated in Part II of the Reassessment[1] that addresses critical considerations in the application of the toxic equivalency factor (TEF) method (Part II, Chapter 9, section 9.2.6, p. 9-10). In addition, conclusions in Part III appear to be congruent with discussions in Part II, Chapter 9, and in the Reassessment overall. No major omissions were identified in the Reassessment, but several aspects need to be addressed or updated.

DIOXIN-LIKE COMPOUNDS

The compounds that are the focus of the Reassessment include 7 of 75 polychlorinated dibenzo-*p*-dioxins (PCDDs), 10 of 135 polychlorinated dibenzofurans (PCDFs), and of the total 209 polychlorinated biphenyls (PCBs), only 4 of the 12[2] previously defined as 2,3,7,8-tetrachlorodibenzo-

[1]*The Exposure and Human Health Reassessment of 2,3,7,8-Tetrachlorodibenzo-p-dioxin (TCDD) and Related Compounds* (EPA 2003a, Part I; 2003b, Part II; 2003c, Part III) is collectively referred to as the Reassessment.

[2]Of the 12 PCBs that received TEF values from the World Health Organization (WHO), the U.S. Environmental Protection Agency (EPA) only considered four in the Reassessment. The remaining eight mono-ortho-substituted PCBs were not considered at this time because of concerns about the accuracy of previous in vivo and in vitro toxicological (relative potency) results, given that a recent study found that many preparations of "pure" mono-ortho PCBs actually contained potent dioxin-like coplanar PCBs as minor contaminants.

p-dioxin (TCDD)-like by the World Health Organization (WHO) (van den Berg et al. 1998). The toxic potency of each of these DLCs (their TEFs) is expressed relative to that of TCDD (also referred to as dioxin), the most potent member of this chemical class (Part III, Table 1.3, p. 1-20). These chemicals are classified as DLCs, given their similarity in chemical structure and physiochemical properties, their ability to invoke a common battery of toxic responses by a common aromatic hydrocarbon receptor (AHR)–dependent mechanism in vivo, and their ability to be persistent environmentally and to bioaccumulate. The lack of inclusion of the eight mono-ortho PCBs previously assigned TEFs by WHO in the Reassessment at the present time is due to concerns that the previously reported activity of many of these chemicals might have been primarily or partially due to a dioxin-like PCB contaminant (PCB126) present in these mono-ortho PCB preparations (DeVito et al. 2003). Although some AHR-dependent toxic effects have been observed with mono-ortho PCBs prepared by methods that should not produce the more toxic DLCs, it remains to be determined whether most of the reported toxicological effects and resulting relative potency (REP) values of these chemicals are due to contaminants, mono-ortho PCBs, or both. Given this uncertainty and the fact that reanalysis of the mono-ortho PCBs as pure compounds is currently being reexamined, they were not included in the list of relevant DLCs for consideration in the Reassessment. Once these issues are resolved, the mono-ortho PCBs should be considered in a follow-up to the Reassessment if they are documented to produce AHR-dependent toxic effects.

MAJOR ISSUES, ASSUMPTIONS, AND UNCERTAINTIES

The relative toxicological and biological potency of a complex mixture is assessed by the TEF approach. Current TEFs are "order-of-magnitude" qualitative values for dioxins, other than TCDD, and DLCs that were established by a WHO expert scientific panel that examined a large scientific database of REP estimates from in vivo and in vitro studies of the biochemical and toxic effects. In the TEF approach, the concentration of the individual compound present in the mixture (determined by instrumental analysis) are multiplied by their specific TEF value, and the sum is expressed as the TCDD toxic equivalent quotient (TEQ). Summation of the calculated TEQs for all active TCDD-related compounds in a sample extract yields the total TEQ for the specific sample extract. Numerous assumptions underlie the use of the TEF/TEQ approach; these have been well delineated, and the major aspects are discussed in detail in the Reassessment (Part II, Chapter 9, and Part III, Chapter 1, section 1.2). These assumptions and uncertainties are described and discussed below.

Role of AHR

Assumption: AHR mediates most toxicities produced by TCDD and other PCDDs, PCDFs, and coplanar PCBs that are AHR agonists. Although AHR is necessary, the ability of TCDD, other dioxins, and DLCs to produce their biochemical and toxicological effects results from downstream events regulated by AHR and AHR-dependent gene expression. The role of AHR in the toxic and biological effects of the TCDD, other dioxins, and DLCs has been supported by a substantial number of quantitative structure-activity relationship, biochemical, genetic, and targeted *Ahr* knockout studies.

AHR-Independent Mechanisms Excluded

Assumption: Effects mediated by other mechanisms (AHR independent) and interactions with other chemicals are ignored. AHR-independent effects of TCDD have been previously observed, including effects on intracellular calcium levels (Puga et al. 1997), changes in gene expression (Oikawa et al. 2001), and selected toxicity in *Ahr* knockout mice (Fernandez-Salguero et al. 1996; Lin et al. 2001). Whether all TCDD-related compounds produce these effects is unknown. Although these mechanisms may play a role in the biochemical effects of TCDD, other dioxins, and DLCs, their significance and role in the overall toxic effects of these compounds remain to be established. However, the Reassessment should acknowledge that AHR-independent effects of TCDD occur and that future studies might demonstrate a role for these effects in the overall toxic and biological effects of TCDD, other dioxins, and DLCs.

Uncertainty of TEF Values

Considering the uncertainty in selection of the TEFs and the information presented on REPs and TEFs in the Reassessment, the 2000 EPA Science Advisory Board (SAB) Panel "questioned whether the uncertainty in the TEFs and the application of this approach to predicting risks due to current levels of exposure was adequately presented" (EPA SAB 2001, p. 29). They concluded that the Reassessment should acknowledge the need for better uncertainty analysis of the TEF values, and although no current method for doing so has been endorsed by the scientific community, several approaches were suggested, such as the use of probabilistic distributions of TEF values in TEQ evaluation (Finley et al. 2003). Available information indicates a considerable amount of variability in the REP value data that were used to derive the WHO TEF values. In addition, although the WHO TEFs were derived based on a scientific consensus evaluation of the available REP values using defined weighted criteria for individual studies, details of the quantitative basis of this weighting scheme were not clearly presented in the description publication (van den Berg et al. 1998). These issues would contribute to variability and uncertainty in the application of the WHO TEF values to health risk assessment. Application of a mathematical value or percentage of the overall range of REP values, such as those described by Finley et al. (2003), would be one way to make the process of determining the specific TEFs more transparent and to provide a standard method to develop TEFs for other TCDD-related compounds that may be added at a later date. Some members of the 2000 EPA SAB Panel also recommended "that, as a follow up to the Reassessment, EPA should establish a task force to build 'consensus probability density functions' for the thirty chemicals for which TEFs have been established, or to examine related approaches such as those based on fuzzy logic" (EPA SAB 2001, p. 29). The committee strongly recommends that the EPA consider inclusion of uncertainty analysis of the TEF values as a follow-up to the current Reassessment.

Consistency of DLC REP Values

Assumption: The REP of a chemical in this group is presumed to be equivalent for all end points of concern and for all exposure scenarios, and all are full agonists. Although most in vitro and in vivo studies support this assumption, the 2000 EPA SAB Panel noted in their review of the Reassessment (EPA SAB 2001) that there are reports of significant differences between the potency of some dioxins, other than TCDD, and some DLCs and specific "toxic end points" is illustrated in Table 5-4 and Table 2-4 in the Integrated Summary (SAB 2001, p. 31). For example, the panel indicated that "1,2,3,7,8- PeCDF (pentachlorodibenzofuran) has the same tumorigenicity as TCDD but was ~38 times weaker for teratogenicity; the other congener, 2,3,4,7,8-PeCDF had half the tumorigenic potency as TCDD, but is ~8 times less potent for teratogenicity" (EPA SAB 2001, p. 31). However, although it was noted that no other examples of that difference were presented in the Reassessment, the observations did raise some concerns about whether all toxic end points could be combined into a single TEF value. The 2000 EPA SAB Panel suggested that "because TEFs vary among different endpoints as well as congeners, it would also be helpful for the document to note that, as data becomes available, it may be possible to derive TEQs [and TEFs] for different endpoints" (EPA SAB 2001, p. 31). The committee agrees that end-point-specific TEFs should be used in those situations in which one is interested in assessing the effects of a sample on a specific end point; however, for general monitoring or screening approaches (that is, for TCDD-related compounds in food and environmental samples) in which all end points should be considered, TEF values that are based on all end points should be used.

Use of TEFs for DLC Body Burdens

Perhaps the issue of greatest concern in this section of the Reassessment is whether the current WHO TEFs, which were developed to assess the relative toxic potency of a mixture to which an animal is directly exposed by dietary intake, are appropriate for the assessment of internal TEQ concentrations and potential toxic effects. Application of the equation relating body burden, half-life, and bioavailability to congeners other than TCDD to give TCDD equivalents based on intake TEF values assumes that the TEF allows adequately for any difference between the congener and TCDD for half-life and bioavailability aspects. In addition, if exposure and estimated body burden of dioxins, other than TCDD, and DLCs are based on measured tissue concentrations, then converting the tissue concentration to a TEQ with TEFs derived from external doses might not be appropriate and might introduce significant uncertainties into the total TEQ estimate. In fact, previous studies have suggested that, because of toxicokinetic differences, the REP values for three PCDFs (2,3,7,8-tetrachlorodibenzofuran [TCDF], 1,2,3,7,8-PeCDF, and octachlorodibenzofuran [OCDF]) were greater when estimated from tissue concentration than when estimated from administered dose (DeVito et al. 1997). These data would support development of body burden TEF values in which the level of toxicity is directly related to body burden concentrations of a given DLC. Questions have also been raised about including octachlorodibenzo-*p*-dioxin (OCDD) and OCDF in the TEF scheme. Differences in the toxicokinetics of these compounds from other chemicals complicated early studies. OCDF and OCDD were originally assigned a TEF of zero because they failed to produce effects in early toxicity studies. However, both OCDF and OCDD are poorly absorbed in the gastrointestinal tract (Birnbaum and Couture 1988; DeVito et al. 1998) and significant TCDD-like effects of each were observed only after repeated doses were given over an extended time to allow accumulation in tissue (Couture et al. 1988; DeVito et al. 1997). Whether toxicokinetic differences of dioxins, other than TCDD, and DLCs exist that would similarly affect their REP and thus their TEFs need to be determined. However, these results raise concerns about the use of intake TEFs for body burden TEQ determinations and suggest that, if possible, it would be more appropriate to generate an additional set of TEFs for body burden tissue equivalents that could be used for DLC risk evaluation purposes. In addition, the use of intake TEFs for body burden TEQ determinations questions the overall conclusion that TCDD, other dioxins, and DLC body burden in humans is currently close to levels that reportedly produce adverse effects in animals. Would it be higher or lower depending on the specific TEFs applied? A discussion of this point could not be found in the sections on toxic equivalents and should be included.

Additivity of DLCs

Assumption: Mixtures exhibit additive toxicities based on TEFs of individual chemicals. Additivity is a particularly critical assumption for the TEF approach. Considerable discussion of this issue is provided in the Assessment, Part II, Chapter 9, and from an overall perspective, this assumption appears valid, at least in the context of risk assessment. Additivity in biochemical and toxic responses by the indicated compound has been supported by numerous controlled mixture studies in vitro and in vivo and is scientifically justifiable. That support is not the case with other non-DLC PCDDs, PCDFs, and PCBs that are reported to be partial agonists or antagonists. The presence of partial agonists or antagonists in a complex mixture or in vivo would likely reduce the overall toxic potency (TEQ) of a mixture when tested in an animal when compared with the TEQ potency calculated simply from application of TEF values to individual compounds measured by instrumental analysis of the mixture. In fact, the ability of some non-DLC PCBs and PCDFs to inhibit TCDD-induced cytochrome 4501A1 protein (CYP1A1) activity and immunotoxicity in C57BL/6J mice has been reported (Bannister et al. 1987; Davis and Safe 1988; Biegel et al. 1989; Chen and Bunce 2004), as has the ability of a lower-affinity synthetic PCDF, such as 6-methyl-1,3,8-trichlorodibenzofuran (6-MCDF), to inhibit TCDD-induced CYP1A1, teratogenicity, immunotoxicity, and porphyria in rodent models in vivo (Astroff et al. 1988; Harris et al. 1989; Bannister et al. 1989; Yao and Safe 1989). These studies indicate that persistent non-DLCs can affect the magnitude of toxic and biological effects produced by a defined amount of TEQ calculated for a given complex mixture. However, given that the presence and concentration of these chemicals in a particular extract can vary dramatically and that very few published studies demonstrate significant alterations in the additive toxicities of dioxins, other than TCDD, and DLCs by other persistent non-DLC AHR ligands in vivo, the assumption of additivity of dioxins, other than TCDD, and DLCs should be considered a valid approach at the present time. Several published papers have demonstrated synergistic activation of AHR-dependent gene-expression effects that involve cross-talk between signaling pathways even at low concentrations. However, with respect to AHR-dependent toxic effects, current data are consistent with ligand and agonist additivity, which is a key assumption of the TEF/TEQ approach. However, EPA should acknowledge the possibility that the presence of non-DLC AHR antagonists in a complex mixture could affect the magnitude and overall toxic effects produced by the calculated amount of TEQs present in a mixture containing such compounds.

Rodent-to-Human Prediction

Assumption: REP of TCDD, other dioxins, and DLCs in rodent models is predictive of REP in humans, given that the rank-order potency is similar between species. Results from available in vivo, in vitro, and accidental and occupational exposure studies are generally consistent with this assumption. Numerous investigators have reported species-specific differences in AHR ligand binding affinity of TCDD, other dioxins, and DLCs. Depending on the system examined, the estimated affinity of binding of TCDD (and related compounds) to the human AHR is about 10-fold lower than that observed to the AHR from "responsive" rodent species and is comparable to that observed to the AHR from "nonresponsive" mouse strains (Roberts et al. 1990; Ema et al. 1994; Poland et al. 1994; Ramadoss and Perdew 2004). This reduced affinity appears to be at least in part due to a single amino acid substitution within the ligand binding domain of the human and "nonresponsive" mouse AHRs (Ema et al. 1994; Poland et al. 1994; Ramadoss and Perdew 2004). Although the affinity of binding of TCDD and related compounds to the human AHR is reduced compared with rodent AHRs, the qualitative and quantitative rank-order potency of these chemicals is similar. In addition to ligand binding, the REP of TCDD and related compounds to induce AHR-dependent gene expression in human cells is also reduced by up to 10-fold (Roberts et al. 1990; Harper et al. 1991; Xu et al. 2000; Zhang et al. 2003; Peters et al. 2004; Silkworth et al. 2005). Because TEFs are expressed relative to the toxicity of TCDD, the shift in TEF values of dioxins, other than TCDD, and DLCs appears to be similar between species. Several recent papers have reported that biological and toxicological responsiveness of humans to TCDD, other dioxins, and DLCs can vary up to 10-fold in vivo and in vitro and that these interindividual differences in responsiveness are not due to specific polymorphisms in AHR (Anttila et al. 2000; Harper et al. 2002; Cauchi et al. 2003). Not only do the documented species differences in AHR ligand binding and AHR responsiveness need to be addressed or taken into consideration with regard to rodent-to-human extrapolation, but the issue of interindividual variability among humans in their responses to TCDDs, other dioxins, and DLCs also needs to be considered when assessing human risk. The rank-order potency of other non-DLC AHR agonists is not necessarily similar between species, and if these chemicals are to be included in the TEF methodology in the future, species-specific TEFs would need to be developed.

Other Persistent AHR Agonists

Assumptions: Although other classes of persistent halogenated environmental chemicals that are structurally related to TCDDs, other dioxins, and *DLCs have been identified, they are excluded because there are limited toxicological data and no validated TEFs for these chemicals.* Another important source of uncertainty is the acknowledged likelihood that other persistent halogenated chemicals, such as brominated and mixed chloro and bromo coplanar chemicals, are present in environmental mixtures, the identities of which are just now emerging and for which TEFs have not yet been established (Reassessment, Part II, section 9.3.5; Part III, section 1.1). Many of these chemicals have been examined and observed to produce adverse AHR-dependent effects in vivo (Birnbaum et al. 1991, 2003). In fact, one mixed polychlorinated and polybrominated dibenzo-*p*-dioxin (2,3-dichloro-7,8-dibromo-dibenzo-*p*-dioxin) produced AHR-dependent toxicity in vivo (wasting and thymic involution) at concentrations up to 10 times lower than that of 2,3,7,8-TCDD (IPCS 1998a, p. 879, Table 50). Although significant information on the polybrominated dibenzo-*p*-dioxins and furans (PBDDs and PBDFs) is available and REP values for some of these compounds have been developed, there still are few toxicological and environmental distribution studies on these compounds. However, IPCS (1998a) suggested that development of TEFs for selected PBDDs and PBDFs is justified given their existing similarities in structure, mechanism, and potency to PCDDs and PBDFs. There are also many other classes of polyhalogenated chemicals that are known to bind to and activate AHR (polychlorinated naphthalenes, benzenes, azobenzenes, azoxybenzenes, and others), and some of these have also been shown to produce TCDD-like effects. However, the primary issue for the lack of consideration of these other TCDD-related compounds in the current assessment is that insufficient data are available on these chemicals, there are no currently determined or validated REPs and TEFs, and questions remain about the presence and persistence of these chemicals in the environment, food, and organisms. EPA should include these chemicals in the TEQ calculations when validated TEFs are developed.

Natural and Synthetic Non-DLCs AHR Agonists

Assumptions: Synthetic and natural non-DLC AHR agonists with a short biological half-life and lower AHR binding affinity do not interfere with PCDD-, PCDF-, and PCB-dependent TEQ predictions. It has been recognized for several years that human and animal diets contain relatively high concentrations of naturally occurring AHR agonists and antagonists (Denison et al. 2002; Denison and Nagy 2003; Jeuken et al. 2003) and that there are non-dioxin-like halogenated aromatic hydrocarbons (HAHs) (PCBs and PCDFs) that are relatively potent AHR antagonists (described below). From a pharmacological and receptor binding kinetics point of view, if one assumes that the binding of these non-DLC agonists or antagonists to AHR is similar to that of TCDD (that is, binding is essentially irreversible) (Farrell et al. 1987; Bradfield and Poland 1988; Henry and Gasiewicz 1993; Brown et al. 1994; Petrulis and Bunce 2000), then the presence of relatively constant and high concentrations of relatively weak non-dioxin-like agonists or antagonists in blood and tissue (e.g., from chronic consumption of relatively high levels of these chemicals) could be expected to produce AHR-dependent effects or inhibit the overall toxic and biological effects produced by a defined amount of TEQ calculated from TCDD-related compounds present in a sample extract.

In most published studies, these metabolically labile non-DLC AHR agonists do not produce AHR-dependent toxicity; however, a few studies have reported the ability of some of these chemicals to produce TCDD-like toxic effects. b-Naphthoflavone (a polycyclic aromatic hydrocarbon [PAH] AHR agonist) was reported to produce thymic involution and splenomegaly in "AHR-responsive" C57 but not "AHR-nonresponsive" DBA mice (Silkworth et al. 1984) as well as wasting and brain developmental effects in fish (Grady et al. 1992; Dong et al. 2002). Developmental exposure of rats to indole-3-carbinol (I3C), a naturally occurring AHR ligand that can be converted in acidic conditions in the stomach into potent AHR agonists, including the high-affinity AHR agonist indolo-[3,2b]-carbazole (ICZ), was reported to produce some AHR-dependent reproductive effects similar to those of TCDD, although other distinct effects of ICZ were noted (Wilker et al. 1996). In addition, inhibition of cytochrome P450-dependent metabolism of PAHs was reported to result in dioxin-like effects in developing fish embryos exposed to PAHs that are AHR agonists (Wassenberg and Di Giulio 2004a,b). Not only would inhibition of CYP-dependent metabolism increase the persistence of the PAH in fish in vivo, but this scenario could also occur in the environment where organisms are exposed to complex chemical mixtures. In contrast to the above studies, the naturally occurring AHR ligand I3C failed to produce adverse effects in rats not only in a 1-year dietary chronic exposure study (Leibelt et al. 2003) but also in a high-dose, short-term study with subcutaneously administered ICZ for up to 10 days (Pohjanvirta et al. 2002).

The ability of metabolically labile phytochemicals to induce or inhibit induction of CYP1A1-dependent activities by TCDD in cell culture model systems has been reported by numerous laboratories (Williams et al. 2000; Amakura et al. 2002; Jeuken et al. 2003; Zhang et al. 2003). Moreover, while the naturally occurring AHR ligands I3C and diindolymethane have been reported to inhibit TCDD-dependent induction of CYP1A1 in B6C3F1 mice in vivo (Chen et al. 1995, 1996), ICZ failed to interfere with the effects of TCDD in a high-dose 10-day study (Pohjanvirta et al. 2002). Lower-affinity synthetic non-dioxin-like AHR agonists, such as 6-MCDF, have been observed to inhibit TCDD-induced CYP1A1, teratogenicity, immunotoxicity, and porphyria in rodent models in vivo (Astroff et al. 1988; Bannister et al. 1989; Harris et al. 1989; Yao and Safe 1989). The ability of some non-dioxin-like PCBs and PCDFs to inhibit TCDD-induced CYP1A1 activity and immunotoxicity in C57BL/6J mice has also been reported (Bannister et al. 1987; Davis and Safe 1988; Biegel et al. 1989; Chen and Bunce 2004). In addition, administration of a synthetic flavonoid antagonist of the AHR (3'-methoxy-4' nitroflavone) to transgenic mice was observed to inhibit TCDD-inducible CYP1A1 and an AHR-responsive β-galactosidase transgene (Nazarenko et al. 2001).

In EPA's Reassessment, a strong case is made for the distinctiveness of highly persistent AHR agonists, versus readily metabolized ones, in terms of toxicological responses and risk assessment. However, the limitation with regard to the lack of knowledge of the effects of the large number of naturally occurring and synthetic AHR ligands on the overall toxic potency of TCDD-related compounds was acknowledged in the Reassessment (Part III, p. 9-40, lines 27 to 28). Although few studies have examined the effects of non-DLC AHR agonists or antagonists on the overall toxic and biological potency of TCDD-related compounds, a few in vivo studies do provide supporting evidence that metabolically labile AHR agonists or antagonists can actually reduce the overall toxic potency of TCDD and presumably other dioxins and DLCs. On the other hand, an excellent correlation between the predicted TEQ and the magnitude of the observed response was observed in several studies examining the effects of real-world samples (soot, incinerator fly ash, sediment leachate, and fish or fish extracts) in animals exposed to these samples in vivo (DeCaprio et al. 1986; Silkworth et al. 1989; Suter-Hofmann and Schlatter 1989; Tillitt and Wright 1997; Powell et al. 1997). While the occurrence of AHR-dependent antagonism by phytochemicals and other AHR antagonists in humans has yet to be confirmed, given species similarities in the AHR and AHR signaling pathway and the relatively high concentrations of many naturally occurring dietary AHR antagonists, the possibility remains that interactions or interferences between natural AHR agonists and TCDD-related compounds might occur. Non-DLC AHR agonists could affect the TCDD-related compounds dose-response relationships for short biological responses (that is, gene induction) and contribute to an additive response for the end points. However, the metabolic lability (that is, lack of persistence) of these compounds prevent them from affecting longer-term dose-response relationships (including threshold and nonlinear assumptions) for toxic end points, such as cancer. That is one reason for the Reassessment to focus only on TCDD, other dioxins, and DLCs that are documented to produce AHR-dependent toxicity. Although these interactions would not affect individual TEF values or the calculation of an overall TEQ determined in controlled laboratory experiments, they could affect the magnitude and overall toxic effects produced by a defined amount of total TEQs calculated from intake or present in the body. Accordingly, EPA should acknowledge in the Reassessment the potential for non-DLCs to affect the overall biological and toxic potency of a defined amount of TEQs present in a complex mixture of chemicals and propose considering these compounds in the overall calculations when and if sufficient and appropriate in vivo data become available in the published literature to support their modulatory effect on DLC- and AHR-dependent toxicity.

KEY STUDIES AND PUBLICATIONS TO BE INCLUDED

Several relatively recent studies not included in the Reassessment support using the TEF/TEQ approach for noncancer and cancer end points; their inclusion would greatly strengthen the Reassessment.

- Studies in rats with TCDD or heptachlorodibenzo-*p*-dioxin (HpCDD) revealed that the REP derived from acute toxicity studies were the same as that obtained in a subchronic and chronic toxicity study; both had a TEF of ~0.007 for HpCDD, although no confidence bounds were provided (WHO TEF = 0.01) (Viluksela et al. 1997a).
- A mixture of four PCDDs or individual PCDDs at equipotent doses (based on TEFs) to rats produced comparable biochemical changes after single as well as multiple doses. The authors concluded that TEFs from acute toxicity studies can accurately predict the toxicity of dioxins, other than TCDD, and DLC mixtures regardless of whether they are administered as single compounds or as a mixture, the results supporting additive toxicity for those compounds (Stahl et al. 1992; Viluksela et al. 1998a,b).
- Rats given a mixture of two PCDDs, four PCDFs, and two PCBs (in a ratio found in foodstuffs) at a concentration of 2.0 μg TEQ/kg of body weight produced adverse reproductive and developmental effects comparable to those at a TCDD concentration of 1 μg/kg (Hamm et al. 2003). The authors concluded that the TEQ approach was a reasonable predictor of the reproductive effects studied.
- Application of TEFs adequately predicted the increased incidence of liver tumors in rats (hepatocellular carcinoma and cholangiocarcinoma) induced by exposure to a mixture of TCDD, 3,3',4,4',5-PCB, and 2,3,4,7,8-

PeCDF compared with an equivalent concentration of TCDD (Walker et al. 2005).

CONCLUSION AND RECOMMENDATIONS

Overall Conclusion

Overall, even given the inherent uncertainties and limitations, the TEF method, when applied correctly, is a reasonable, scientifically justifiable, and widely accepted method to estimate the relative toxic potency of dioxins, other than TCDD, and DLCs on human and animal health.

Specific Conclusions and Recommendations

- *AHR-independent mechanisms excluded.* AHR-independent effects of TCDD have been reported, and although their significance and role in the overall toxic effects remain to be established, the Reassessment should acknowledge the existence of these AHR-independent effects because future studies may demonstrate that they play some role in the overall toxic and biological effects of TCDD, other dioxins, and DLCs.
- *Uncertainty of TEF values.* A significant degree of uncertainty exists in the current consensus TEFs, and the quantitative weighting considerations that have gone into their establishment are not clear. While the Reassessment should acknowledge the need for better uncertainty analysis of the TEF values, extensive and appropriate uncertainty analysis would take considerable time and effort. Accordingly, the committee endorses the recommendation of some members of the 2000 EPA SAB Panel "that, as a follow up to the Reassessment, the EPA should establish a task force to build 'consensus probability density functions' for the thirty chemicals for which TEFs have been established, or to examine related approaches such as those based on fuzzy logic" (EPA SAB 2001, p. 29).
- *Consistency of REP values.* Most in vitro and in vivo studies support the assumption that the indicated dioxins, other than TCDD, and DLCs are not only full agonists but that their REP is similar for all end points of concern and exposure scenarios. However, significant end-point-specific differences in the REP of some dioxins, other than TCDD, and DLCs have been reported and whether other differences exist remains to be determined. Consistent with the recommendations of the 2000 EPA SAB Panel, this committee also suggests that it would be appropriate for the Reassessment to note that end-point-specific TEFs/TEQs might be derived as data become available and that those specific values be used when that end point is being considered. It should also be made clear that general monitoring or screening approaches (that is, for TCDD-related compounds in food and

environmental samples) should use TEF values that are based on REPs values of all end points.

• *Use of TEFs for DLC body burdens.* This is perhaps the greatest issue of concern in this section of the Reassessment because it remains to be determined whether the current WHO TEFs, which were developed to assess the relative toxic potency of a mixture to which an animal is directly exposed by dietary intake, are appropriate for the assessment of internal TEQ concentrations and potential toxic effects. The issue was not well described or well justified in the Reassessment and might be incorrect. It is further complicated by an EPA paper (DeVito et al. 1997) suggesting that use of TEFs for DLC body burdens might not be appropriate for some PCDFs. The issue would be further complicated if toxicokinetic differences of other DLCs similarly affect their REP. Overall, it remains to be determined whether intake TEFs are appropriate for body burden TEQ determinations. If body burdens are going to be used as the dose metric, the committee recommends that a separate set of body burden TEFs be developed and applied for this evaluation or that the appropriateness of intake TEFs for body burden TEQs be scientifically justified. Without these corrected values, the overall TEQs estimated by use of intake TEFs could be inaccurate.

• *Role of AHR and additivity of DLCs.* These aspects are well described and well supported by extensive numbers of scientific studies. However, EPA should acknowledge the possibility that AHR antagonists present in a complex mixture could affect the magnitude and overall toxic effects produced by a calculated amount of total TEQs present in a given sample even if they do not affect the TEQ calculations. This issue was not addressed in the Reassessment.

• *Rodent-to-human prediction.* Although the REP of dioxins, other than TCDD, and DLCs in rodent models is predictive of REP in humans from a qualitative rank-order potency point of view, some species-specific differences in AHR ligand binding affinity of TCDD, other dioxins, and DLCs have been observed. However, because TEF values are expressed relative to that of TCDD in the individual species, the TEF values for dioxins, other than TCDD, and DLCs appear to be similar between species. If significant differences in the REP of dioxins, other than TCDD, and DLCs are found between humans and other species, then adjustments should be made in the TEFs, and these should be acknowledged in the Reassessment.

• *Other AHR agonists.*

— *Related HAH DLCs.* Lack of consideration of other persistent halogenated chemicals, such as brominated, chlorinated, and mixed chloro and bromo coplanar chemicals, which clearly exert their toxic and biological effects in an AHR-dependent manner could result in underestimation of the overall TEQ for a given sample. Although REP values and TEFs have been developed for some of these chemicals, few studies have been carried out with most of them, and their relative toxic potency is unknown. Given the structural similarities and mechanism of action of these chemicals in vivo and in vitro with the established compounds, as validated REP values become available, TEFs should be assigned, and these chemicals should be included in the TEF/TEQ approach. This course of action should be noted in the Reassessment.

— *Synthetic and naturally occurring non-DLC AHR ligands.* A large number of synthetic and naturally occurring non-DLC AHR ligands have been identified and are present in human diets and presumably in blood and tissues. The assumption that non-DLC AHR agonists with a short biological half-life do not interfere with DLC-dependent TEQ predictions for mixtures is controversial and remains to be confirmed. Although receptor binding kinetic evaluations suggest that these chemicals could interfere with TCDD, other dioxins, and DLCs if at high concentrations in blood and tissue, few of these metabolically labile non-DLC AHR agonists have been observed to directly produce AHR-dependent toxicity. The Reassessment makes a strong case for the ability of only highly persistent AHR agonists to produce toxicity, but the lack of knowledge of the effects exerted by the large number of naturally occurring dietary and synthetic AHR ligands on the overall toxic potency of TCDD, other dioxins, and DLCs still leaves the question open, particularly with regard to humans. Although these AHR ligands would not affect TEQ calculations, they could affect the magnitude of the toxic and biological effect of a defined amount of TEQ. This point should at least be made clear in the Reassessment, and when a sufficient number of published studies demonstrate the ability of non-DLC AHR agonists or antagonists to modulate the overall effects of DLCs, then EPA should consider how these chemicals would affect the current TEF/TEQ approach for potency estimates.

- *WHO's plan to reexamine DLC TEFs in 2006.* The major issues of concern described above for the TEF approach will also be the focus of a meeting of the International Programme on Chemical Safety (announcement in IPCS 2004). The issues include (1) considering methods and approaches for deriving TEFs, including quantitative (statistical) methods, such as establishing an uncertainty range of available REP data and application of a specified cut-off value to derive TEF values, application of weighting factors to existing data, and related issues; (2) determining whether to continue to include mono-ortho PCBs in the present TEF concept; (3) considering whether other compounds should be considered for inclusion in the TEF concept, taking into account the prerequisites for inclusion outlined by Van den Berg et al. (1998); and (4) determining the applicability of the use of TEFs to estimate intake versus internal concentrations and to what extent could or should internal WHO TEF factors be established in the future? EPA should consider the outcome of the IPCS TEF update meeting and incorporate the issues and changes into the Reassessment.

• *Updating the Reassessment.* Although the Reassessment clearly states that the WHO TEFs of 1998 will be used for assessment and calculation, if or when TEF values are changed or new chemical TEFs are added by the current or future WHO TEF panels (such as the 2005 panel), EPA should consider incorporating the new TEF values and methods for TEQ determination.

4

Exposure Assessment

The Reassessment[1] addresses exposure in terms of sources, environmental fate, environmental media concentrations, food concentrations, background exposures, and potentially highly exposed populations including important developmental stages. In this chapter, the committee discusses the exposure characterization section provided in the Reassessment, Part III. Part I of the Reassessment has a wealth of supporting information and comprises an executive summary and three volumes: *Sources of Dioxin-like Compounds in the United States*[2]; *Properties, Environmental Levels, and Background Exposures*; and *Site-Specific Assessment Procedures*.

ASSESSMENT PROCEDURES

The comments in this chapter are directed specifically at the use of exposure assessment in the risk assessment provided in Part III of the Reassessment, but the committee consulted the more detailed companion documents in Part I for supporting information.

---

[1] *The Exposure and Human Health Reassessment of 2,3,7,8-Tetrachlorodibenzo-p-dioxin (TCDD) and Related Compounds* (EPA 2003a, Part I; 2003b, Part II; 2003c, Part III) is collectively referred to as the Reassessment.

[2] Although EPA gave this document the title *Sources of Dioxin-like Compounds in the United States,* it provides information on sources of TCDD, other dioxins, and dioxin-like compounds.

Similar to the Reassessment, Part III, the chapter here is organized into sections on sources, environmental fate, environmental media and food, background exposures, and potentially highly exposed populations and sensitive populations. This chapter has three major sections: an overview and commentary on all aspects of the dioxin exposure assessment with an effort to point out strengths, limitations, and omissions; the committee's findings; and specific recommendations.

OVERVIEW AND COMMENTARY ON EPA'S EXPOSURE CHARACTERIZATION

In this section, the committee provides summary and commentary on key issues related to exposure characterization for 2,3,7,8-tetrachlorodibenzo-*p*-dioxin (TCDD, also referred to as dioxin), other dioxins, and dioxin-like compounds (DLCs). This information includes sources, environmental fate, environmental media and food concentrations, background exposures, and potentially highly exposed populations and particularly sensitive developmental stages.

For sources and environmental fate, EPA had a clearly articulated stepwise approach that the committee primarily accepted with some commentary. The other steps in the exposure assessment are not as easy to track, summarize, and critique. To comment on these steps, the committee used a format that went beyond the simple narrative.

Sources

Summary of the EPA Approach

The type, geographic distribution, and time history of the sources and associated emission magnitudes of TCDD, other dioxins, and DLCs are essential inputs for risk characterization. In Part III of the Reassessment, EPA discusses sources and emissions estimates for 1987 and 1995. More recently, EPA issued a report that includes the year 2000 update on sources and emissions estimates (EPA 2005b). These reports consider emissions of polychlorinated dibenzo-*p*-dioxin (PCDD) and polychlorinated dibenzofuran (PCDF) compounds and dioxin-like polychlorinated biphenyl (PCB) compounds. PCDDs and PCDFs have never been intentionally produced outside research laboratories. They are released to the environment as unintended by-products from various combustion, industrial, and biological processes. PCBs have been produced commercially in large quantities in the United States and other industrialized countries but are no longer commercially produced in the United States and Europe.

Sources of TCDD, other dioxins, and DLCs considered in the Reassessment include combustion sources; metals smelting, refining, and processing industries; and chemical manufacturing, biological and photochemical processing, and reservoir sources. PCDDs and PCDFs are formed in most combustion systems—waste incineration and burning of coal, wood, and petroleum products; other high-temperature sources (such as cement kilns); and poorly or uncontrolled combustion sources (such as forest fires, building fires, and open burning of wastes). PCDDs and PCDFs can be formed during various types of primary and secondary metals operations, including iron ore sintering, steel production, and scrap metal recovery. PCDDs and PCDFs can be formed as by-products from the manufacture of chlorine-bleached wood pulp, chlorinated phenols (e.g., pentachlorophenol [PCP]), PCBs, phenoxy herbicides (e.g., 2,4,5-trichlorophenoxyacetic acid, or 2,4,5-T), and chlorinated aliphatic compounds. Recent studies suggest that PCDDs and PCDFs can be formed under certain environmental conditions (e.g., composting) from the action of microorganisms on chlorinated phenolic compounds. EPA also reported that PCDDs and PCDFs have formed during photolysis of highly chlorinated phenols.

Reservoir sources of TCDD, other dioxins, and DLCs are materials or places that contain previously formed PCDDs and PCDFs or dioxin-like PCBs and have the potential for redistributing and circulating these compounds into the environment. Potential reservoirs include soils, sediments, biota, water, and some anthropogenic materials. Reservoirs become sources when they release compounds to the surrounding environment.

Important Aspects of EPA's Approach, Assumptions, and Findings

The key output of the Reassessment regarding sources is provided in Table 4-2 of the Reassessment, Part III, which summarizes an "inventory" of sources for the United States expressed as toxic equivalent quotients (TEQ). In constructing this table, EPA developed a qualitative confidence-rating scheme in which they used qualitative criteria to assign high-, medium-, or low-confidence ratings to the inventory classes. This table and comparisons of the years 1987, 1995, and 2000 are important inputs to EPA's conclusions about long-term trends in the emissions of TCDD, other dioxins, and DLCs (furans and dioxin-like PCBs). In particular, the committee notes that EPA relied more on emissions estimates than environmental and biological media concentrations as a means of characterizing temporal trends in exposure to TCDD, other dioxins, and DLCs.

EPA's use of the inventory table represents a "bottom-up" approach. EPA compiled a list of all potentially important source categories and provided an estimate of the probable magnitude of emissions from each of these categories. Summing these emissions by categories then provides an overall estimate of current and historical emissions. As noted by EPA, this approach comes with large uncertainties in assigning emission values to each category and may exclude an unknown major category or fail to identify a number of minor categories that together provide large emissions. An alternative "top-down approach" would consider levels of PCDD and PCDF compounds in various environmental media (soils, sediments, and so forth) or biological media (vegetation, tree bark, fish tissues, and so forth) and identify the level of emissions required to account for these PCDD and PCDF levels. The top-down approach uses fate modeling and mass-balance analyses applied to environmental samples along with a number of assumptions to determine the extent to which deposition matches emissions (Rappe 1991; Harrad and Jones 1992; Brzuzy and Hites 1995; Eisenberg et al. 1998). Because of discrepancies among estimates of TCDD and related compounds in reservoirs relative to known sources, several researchers using a top-down approach concluded that EPA estimates of historical national emissions might underestimate emissions (Rappe 1991; Harrad and Jones 1992; Brzuzy and Hites 1995; Eisenberg et al. 1998). This suggests the possibility of unknown sources. Although the bottom-up and top-down approaches come with uncertainties, EPA could benefit substantially from using both approaches simultaneously to set plausible bounds on the historical and current trends in emissions. The committee recognizes that each approach has significant limitations. For example, the identification of ball clay as a potential source represents an interesting case, because it represents an identified (and managed) new source. In the absence of any other information, a bottom-up or a top-down approach is unlikely to find a minor contributor, such as ball clay, to overall national-level TEQ.

One of the most important aspects of the EPA analysis emerges in the discussion of the trends over time. With the most recent update to the inventory (EPA 2005b), there were dramatic declines relative to 1995 and 1987 in the emissions of TCDD, other dioxins, and DLCs from identified major sources. Unfortunately, the Reassessment and the background documents do not provide sufficient information for the committee to review the emission inventory table inputs, either the qualitative assessments or the quantitative estimates. In the current organization of the Reassessment, EPA does not clearly lay out the path for derivation of the emissions numbers. The lack of clarity makes a task as basic as checking the calculations and logic difficult.

Environmental Fate

Summary of the EPA Approach

Part III of the Reassessment provides a summary of key findings about the transport and environmental fate of TCDD, other dioxins, and DLCs.

Another apparent purpose of section 4.2 in Part III of the Reassessment is for EPA to make clear that assessment of environmental fate cannot be based on TEQ but must be based on individual congeners, but in section 4.1, EPA presents estimates of environmental releases as TEQ. They elected to present TEQ in place of mass quantities to better facilitate comparisons across sources. For purposes of environmental fate modeling, however, EPA notes that it is important to use the individual PCDD, PCDF, and PCB congener quantities rather than TEQ, because the physical and chemical properties of individual congeners vary and will behave differently in the environment. This material on the need to address specific congeners appears to have been added to the Reassessment in response to the Science Advisory Board's comment that the original dioxin reassessment report (EPA 1994) implied that emissions expressed as TEQ could be used as source terms for modeling transport, fate, and exposure in risk assessments.

Important Aspects of EPA's Approach, Assumptions, and Findings

In its assessment of environmental fate in the Reassessment, EPA makes the following key findings:

- TCDD, other dioxins, and DLCs are widely distributed in the environment as a result of a number of physical and biological processes.
- Because physical and chemical properties vary substantially among individual congeners, the congeners will behave differently as they are transported through and transformed in the environment. Thus, for purposes of environmental fate modeling, it is important to use the individual PCDD, PCDF, and PCB congener levels rather than TEQ.
- Atmospheric transport and deposition of TCDD, other dioxins, and DLCs are the primary means of their dispersal throughout the environment.
- The two primary pathways for TCDD, other dioxins, and DLCs to enter the ecological food chains and human diet are air-to-plant-to-animal and water-and-sediment-to-fish pathways.

In reviewing these findings, the committee notes that they are supported by the source and exposure information provided in the Reassessment. The committee further notes that EPA missed an opportunity to use data on individual congeners to assess how TEQ changes in time and space. Moreover, many EPA findings on sources, fate, and exposure tend to be drawn from temporal and spatial trends in emissions. EPA did not make full use of exposure media concentration data, particularly food concentration data, to confirm that the space and time trends are reflected in exposure media. EPA missed the opportunity to use emissions data for individual congeners combined with fate modeling to assess the persistence of individual congeners to estimate the persistence of TEQ and the spatial distribution of TEQ. Another issue of interest to the committee is how the reliability of the TEQ estimate becomes more uncertain with time. Because of uncertainty about toxic equivalency factors (TEFs) for the more persistent congeners, such as the hexa, hepta, and octa chlorinated congeners, that tend to dominate the TEQ, the reliability of the TEQ characterization degrades with the resulting accumulation of the more persistent congeners. As a result of not considering this issue, EPA does not yet have the ability to determine when reservoir sources will become significant relative to all anthropogenic sources in characterizing the TEQ of TCDD, other dioxins, and DLCs.

Environmental Media and Food Concentrations

EPA developed estimates of concentrations of TCDD, other dioxins, and DLCs in various environmental media, including foods, using only those studies from locations that they considered as representing background levels of these compounds. The extent to which regions with high exposures were either captured or excluded is not clear in the Reassessment. Moreover, because background has a continuum of low to high concentrations, it is also not clear where the line was drawn to distinguish background from "not background."

Although TCDD, other dioxins, and DLCs in food and environmental media have been declining over the last three decades, the presence of these compounds in foods (primarily in animal fats and oils) now represent 90% or more of human exposure (IOM 2003). However, there are significant uncertainties inherent in calculating with accuracy or precision dietary exposure because of the limited analyses of individual foods; the methodological improvements over time with corresponding lowering of the limits of detection; the limited information on the congener composition of various foods; the values assigned to "nondetects"; the alterations in concentrations of TCDD, other dioxins, and DLCs due to methods of preparation and cooking; the wide diversity of human dietary composition and consumption patterns; and the inherent inaccuracies of the instruments used to assess dietary intake in humans (IOM 2003). The Reassessment extensively details the information available at the time (Part I, Volume 2, Chapters 3 and 4) and briefly mentions the Institute of Medicine (IOM) report (Part III, section 4.3). EPA acknowledges that, in general, the available food data come from "studies that were not designed to estimate national background means" and that "it is not known whether these estimates adequately capture the full national variability" (Part III, section 4.3).

Since the Reassessment, additional studies have estimated human dietary intake of TCDD, other dioxins, and DLCs. In the Netherlands, "the estimated median life-long-averaged intake of the sum of PCDDs, PCDFs, and dioxin-like PCBs in the population is 1.2 pg WHO [World Health Organization] TEQ per kg of body weight per day" (Baars et al. 2004). The estimated median is below the WHO tolerable daily intake of 2 pg TEQ/kg of body weight, and the authors estimated that approximately 8% of the Dutch population have life-long averaged intakes above the WHO tolerable intake level. Charnley and Doull (2005) and the U.S. Food and Drug Administration (FDA) (CFSAN 2005a,b) estimated human food exposures to TCDD, other dioxins, and DLCs between 1999 and 2003 with data derived from the FDA total diet study. These studies provided intake estimates since 2001 in which the average daily intake for all age groups fall below the WHO tolerable daily intake level of 2 pg TEQ/kg of body weight. However the estimates do not include breast-fed infants.

Charnley and Doull (2005) noted that when assessors represent exposure media concentrations of TCDD, other dioxins, and DLCs—primarily in food—below the limit of detection (LOD) by one-half the detection limit, "approximately 5% of the intake estimates for 2-year-olds and 1% of the intake estimates for 6-year-olds exceed the tolerable daily intake by about 10%." When these media concentration measurements below the LOD are set to zero (when only concentration values actually measured are used), "only 1% of intake estimates exceed the tolerable daily intake for 2-year-olds." The committee notes that this reveals the problem of interpreting a "mean" concentration. The arithmetic mean among individuals in these cases is quite sensitive to the treatment of samples below the LOD. One alternative is to avoid the use of sample means and instead consider comparisons based only on percentile concentrations (e.g., median values and 90th percentile individual values). These percentile values only require information about the rank of a sample and thus avoid the impact on central value estimates introduced by LOD assumptions.

In both American (Charnley and Doull 2005) and Dutch (Baars et al. 2004) populations, meat and dairy products account for approximately 50% of the TCDD TEQ consumed in food, but the Dutch consume more TCDD TEQ in fish than do Americans—16% and 5.8%, respectively. Additional data from the U.S. Department of Agriculture Food Safety and Inspection Service confirmed that the contents of TCDD, other dioxins, and DLCs measured in 2002 and 2003 in meat products sold in the United States, including hogs, steer, heifers, young chickens, and young turkeys, have declined significantly from the contents measured in 1994 through 1996, although methodological differences preclude a precise calculation of the decrement (FSIS 2005).

Recognizing that some data gaps will remain in the source inventory, in the environmental media concentrations describing the distribution and environmental fate of these chemicals, and in various parts of the food chain and human tissue concentrations (e.g., breast-milk and serum concentrations), the committee notes that it would be helpful if EPA could set up a congener-specific database of *typical* concentrations in foods for the whole range of PCDDs, PCDFs, and dioxin-like PCBs (those included in the WHO TEF list). Such a database would need to fulfill clear requirements of data quality and traceability (e.g., chemical analysis, representative and targeted sampling, data representative consumer exposure, presentation of data, and handling and presentation of values below the LOD). Making such a database available could improve the transparency of how EPA came to some of the conclusions in the Reassessment. Moreover, if TEF values change, TEQ values can be easily recalculated. Such a database could be updated on a regular basis to evaluate temporal trends. Here, it is important to consider methodological aspects (e.g., reproducibility, sensitivity, specificity of the analytical determinations, inclusion of reference samples, and comparable sampling strategy) to ensure that such a time-trend analysis is useful.

Background Exposures

The section of the Reassessment that addresses background exposures provides a summary of information on human tissue levels, intake estimates, and variability in intake levels.

Tissue Levels

The section of the Reassessment addressing tissue levels evaluates data on concentrations of TCDD, other dioxins, and DLCs in human tissues expressed per gram of lipid and the changes in these concentrations that have occurred in recent decades. The Reassessment acknowledges the difficulty of comparing different data sets because some do not include coplanar PCBs in the estimation of TEQ values. It is clear from the data in Part III, Table 4-5, that TCDD per se is not the main source of TEQs in human lipid. The Reassessment uses the calculation of body burden at steady state along with EPA's associated assumptions given in section 1.3 to calculate the TEQ concentration in human lipid based on the best estimate of current adult intake and the assumption of 25% body fat. The result is about one-half of that actually measured in human lipid. EPA assumes that the discrepancy arises from the presence of an historical body burden and lipid concentration. Given the various assumptions in the estimation of body burden at steady state, especially in relation to application of the TCDD model to congener TEFs, it is reassuring that the TEQ in human lipid predicted by the model are somewhat consistent with the estimated values.

Intake Estimates and Variability in Intake Levels

These sections describe intake estimates and the variability and age-related changes in intake—in particular, by nursing infants.

Potentially Highly Exposed Populations or Developmental Stages

In compiling and evaluating available data on highly exposed populations, EPA considered contamination of food, exposures to workers, and exposures to nursing infants.

In the Reassessment, EPA assumes that contamination incidents in food probably have not and will not lead to disproportionate exposures to populations living near where they occurred. The basis for this assumption is that meat and dairy products in the United States are widely distributed on a national scale. As a result of this assumption, the Reassessment does not comment on any disproportionate exposures due to interaction with contaminated sites.

In considering the distribution of exposures to TCDD, other dioxins, and DLCs in the U.S. population, EPA suggested that variability in exposure probably regresses toward the mean because Americans consume varied diets from multiple sources, meaning that EPA assumed that variations in diet would prevent either very high or very low extremes of exposure. EPA reported that this pooling of the food supply reduces the potential high exposure that could result from high consumption of certain food products. This assumption may be valid, but EPA should provide additional analyses to support it and should also explicitly consider the possibility of populations who violate the assumptions with respect to varied diets and multiple sources (e.g., those who rely on home-produced foods or sustenance fishing). It is of interest that only in the last paragraph of this section is there discussion of measurements reflecting potentially highly exposed groups. Here it is mentioned without further discussion that several European studies showed increased TCDD, other dioxins, and DLC levels in milk and other animal products near combustion sources. EPA did not consider the implications of this finding for the U.S. population. It thus seems that EPA is implicitly assuming that this problem does not exist in the United States.

The Reassessment suggests that no clear evidence demonstrates that increased exposures to TCDD, other dioxins, and DLCs are currently occurring among U.S. workers, but the Reassessment does not document the level of ongoing monitoring and assessment to support this conclusion. Low levels of occupational exposure are not congruent with their reported inventory of sources.

To evaluate the impact of nursing on infants, EPA estimated changes in body burden with a model developed by Lorber and Phillips (2002). This model includes a number of assumptions, including that the fraction of TCDD, other dioxins, and DLCs absorbed by an infant after ingestion is 0.80 and that the dissipation rate of the ingested TEQ is rapid. The developers evaluated the model with the data (from Germany) of Abraham et al. (1998). The EPA evaluation does not necessarily confirm the numerous assumptions (e.g., half-life and uptake). Moreover, the evaluation does not capture variability or uncertainty in the model because of the assumptions. The conclusions at the end of the Reassessment, Part III (p. 4-23, lines 7 to 15), include the presentation of model predictions that are implied to be very precise. Yet, in view of the various assumptions, these results might or might not reflect reality. In light of the amount of supporting information available from other sources, it is unclear why EPA relied primarily on a relatively detailed model with all its inherent uncertainties to report that the annual infant TCDD-TEQ intake from nursing significantly exceeds the currently estimated adult intake of 1 pg TEQ/kg/day. This observation can be easily demonstrated from qualitative findings and simple assessments based on TCDD half-life and lipophilicity, infant body size, breast-milk composition, and breast-milk intake. The committee recommends that EPA consider the value and availability of any data to confirm this modeling result.

COMMITTEE FINDINGS

Is EPA's Exposure Assessment Scientifically Robust?

In preparing its findings, the committee notes that those who will make use of the Reassessment are likely to be interested in issues beyond risk characterization and risk assessment methodology. For example, some users will want to use the Reassessment to decide whether U.S. exposures to TCDD, other dioxins, and DLCs pose an undue health risk, whereas others will want to use the Reassessment to consider alternatives for reducing exposures to these compounds and identifying strategies for achieving reductions of TCDD-TEQ burdens in the U.S. population. In preparing its findings, the committee considered a range of potential uses for the Reassessment—including the following alternatives.

Source Characterization

Clearly, an important opportunity that EPA overlooks is checking the observed decline in overall environmental concentrations against body burden changes over time. For example, the emissions estimates for PCBs and mass-balance evaluation provided recently by Breivik et al. (2002a,b) provide a better opportunity to consider global-scale chemical PCB fate by comparing model results with measured concentrations of PCBs at monitoring stations located in regions of the Northern Hemisphere over the 70-year period from 1930 to 2000. Calculations based on this 70-year estimate of emissions will introduce uncertainties, but such an analysis could build confidence about trends and better inform future investigation.

EPA did not fully address the issue of reservoir sources or explore their potential impacts on the long-term distribution of TCDD, other dioxins, and DLCs as well as the distribution of TCDD TEQ. It also did not fully consider how reservoir effects vary among different congeners and thus cause the TEQ from reservoir sources in soil and sediments to evolve and change in time. Finally, EPA did not address the issue of when reservoir sources are likely to become dominant relative to anthropogenic sources. For example, some studies provide experimental evidence for how TCDD, other dioxins, and DLCs are incorporated in soil and then reemitted (Brzuzy and Hites 1995, 1996; Cousins et al. 1999a,b; Cousins and Mackay 2001; McKone and Bennett 2003).

One of the most important aspects of the analysis emerges in the discussion of the trends over time. Given the importance of properly estimating TEQ and the need for risk analysts to consider the impacts of exposure timing for some potential dose metrics, the EPA inventory should yield estimated TEQs associated with each identified source more transparently. Part III of the Reassessment and the background documents do not provide sufficient information for the committee to review the emissions inventory table inputs, either the qualitative assessments or the quantitative estimates. Although inventories shifted over time with the identification of new sources, EPA did not examine the extent of that shift.

Environmental Fate Assessment

EPA's finding regarding the wide distribution of TCDD, other dioxins, and DLCs is supported by environmental sampling. There have been sufficient measurements to conclude that, as a chemical class, these compounds are widely dispersed in the environment. With regard to individual congeners of PCDDs, PCDFs, and dioxin-like PCBs, a sufficient number of samples are not available to conclude that each individual congener is widely dispersed in the environment.

Although consideration of individual PCDD, PCDF, and PCB congeners would be informative and useful, doing that for more than 200 congeners would be excessive; summing up mass quantities instead of TEQ contributions would be equally bad, and most other inventories (e.g., in Europe and Japan) were also done in TEQs. According to the Reassessment (Part III, p. 1-8), five congeners contribute approximately 80% of the total TEQ in humans: 2,3,7,8-TCDD, 1,2,3,7,8-PCDD, 1,2,3,6,7,8-hexachlorodibenzo-*p*-dioxin, 2,3,4,7,8-

PCDF, and PCB126. Thus, it would be informative to provide congener-specific emissions estimates for these congeners in place of the TEQ estimates.

Environmental Media and Food Concentrations

In reviewing the EPA assessment of environmental media and food concentrations, the committee had the following concerns:

- In using food concentration data to estimate intake, the choice of LOD has significant impact on calculated mean values. EPA was not clear about (1) how it made use of values below the LOD in making intake estimates based on food concentration data, and (2) how its treatment of the LOD had an impact on results. Because the committee found no basis for making recommendations on other aspects of the food intake calculation and because food supply issues are covered in the IOM (2003) report, the committee elected to focus on the LOD issue.
- EPA did not make clear its criteria for distinguishing background from non-background concentrations.
- Relative to dioxin and furan congeners, data on environmental media and food concentrations of dioxin-like PCBs were generally lacking.
- TCDD-TEQ intake estimates from fish consumption did not include direct consumption of fish oils.

The committee finds value in EPA's establishing a congener-specific database of *typical* concentrations for the range of PCDDs, PCDFs, and dioxin-like PCBs (those included in the WHO TEF list). The details of such a database are described above in the overview and commentary section in the subsection on environmental media and food concentrations.

Estimates of Background Exposures

The committee found the text in this section noncontroversial and the conclusions valid. The committee did not find any important errors in this text, but issues arose concerning the interpretation of the background exposure data. It is not clear that the existing database as used by EPA covers all foods consumed by the U.S. population (e.g., data were missing on fish oils). It would be helpful to include or reference in the exposure estimates the most recent data on food intake as produced by FDA. The committee believes that EPA can make more efficient use of the existing data sets on occurrence in foods and on food consumption to assess the distribution of intakes of TEQ for the general U.S. population (different age groups, expressed in picogram per kilogram of body weight per day) as well as intra- and inter-person variability.

Exposures in Highly Exposed Populations or at Key Developmental Stages

In compiling and evaluating available data on highly exposed populations, EPA failed to draw informative conclusions from the numerous studies described in the full document (Part I). Part III of the Reassessment is missing a summary that integrates the information compiled in Part I.

The Reassessment makes statements discounting the potential for having highly exposed groups without clearly documenting the basis for these statements. First, it suggests that, with regard to the commercial food supply, the incidents of contamination by TCDD, other dioxins, and DLCs are likely to be low. Yet the Reassessment provides no formal assessment and almost no data to support that determination. It also states that there is no clear evidence that increased exposures are occurring among U.S. workers. Finally, it reports that no or few studies show evidence of groups in the United States being exposed to highly increased levels of TCDD, other dioxins, and DLCs in situations in which people consume large quantities of foods with high levels of these compounds. In spite of giving substantial attention to nursing infants as a highly exposed group, EPA provides no comment on the potential level of increased exposure that may have arisen during recent contamination episodes involving the commercial food supply (e.g., the ball clay incident and high levels in beef and dairy animals due to PCP-treated wood).

Is There a Clear Delineation of All Substantial Uncertainties and Variabilities?

Overall, the committee finds that EPA has qualitatively identified a number of important uncertainties and variabilities. However, there are some areas for which even the qualitative information provided by EPA was unclear or incomplete. What is more important is that the Reassessment does not quantitatively characterize either variability or uncertainty in exposure except in the limited sense of demonstrating increased average daily dose estimates for children (on a body-weight basis) and analyzing potentially increased exposures for nursing infants during their first few years of childhood.

Source Characterization

The magnitude, type, geographic distribution, and time history of TCDD, other dioxins, and DLC sources are essential components for risk characterization. The interpretation of these factors is an important input to decisions about managing both new and historical (reservoir) sources.

Any errors in interpretation could lead to policies and regulatory actions that are inefficient or ineffective in reducing human exposures to TCDD, other dioxins, and DLCs. EPA exposure characterization excludes basic data quality checks that could provide an opportunity to evaluate key assumptions. The committee notes that EPA did not explore an alternative top-down approach in an effort to evaluate the results from its bottom-up approach for source-to-intake characterization. The Reassessment clearly notes significant uncertainties in estimates of emissions and communicates these uncertainties using qualitative confidence scores (A, B, and C). However, given this clear acknowledgment of significant uncertainties in the emission estimates, the committee questions the reliability of the Reassessment's trend analysis of emissions from 1987 to 2000. EPA does not communicate these uncertainties in the Reassessment's summary and other sections where the trend analysis of emissions is discussed

Environmental Fate Assessment

EPA's finding that atmospheric transport and deposition of TCDD, other dioxins, and DLCs are a primary means of their dispersal throughout the environment is strongly supported by theoretical models in combination with observations of these compounds globally that are more uniform than emissions sources and far from regions of release. However, there is considerable uncertainty about the nature and magnitude of the re-emission process that takes place after deposition.

The EPA finding that the two primary pathways for TCDD, other dioxins, and DLCs to enter the food chain and human diet are air to plant to animal and water and sediment to fish is supported by environmental sampling, but significant uncertainty remains about mechanisms and rates of transfer through food webs. There have been sufficient measurements to conclude that, as a chemical class, TCDD, other dioxins, and DLCs, particularly the more persistent ones, enter humans primarily through animal products and fish. With regard to individual congeners of PCDDs, PCDFs, and dioxin-like PCBs, samples are insufficient to conclude that each individual congener enters humans by these two primary pathways.

The committee concurs with EPA that, although it is appropriate to use TEQ as a metric of release, it must clearly emphasize the uncertainty and limitation of using the TEQ approach.

Environmental Media and Food Concentrations

- It is uncertain whether the existing information on background levels in environmental media and food adequately captures the full national variability.

- The effect of cooking and processing on concentrations of TCDD, other dioxins, and DLCs in foods was considered too limited to draw firm conclusions.
- EPA had only limited access to data that support any conclusions about temporal trends in the occurrence of TCDD, other dioxins, and DLCs in environmental media and foods.

Estimates of Background Exposures

When EPA assumes that less-than-LOD samples equal zero, there are often significant differences between values of TEQ-based estimates of background intake compared with estimates obtained when EPA assumes that less-than-LOD samples equal half or the whole detection limit. This illustrates the importance of analytical method sensitivity in limiting the ability to determine the full range of population variation of PCDDs, PCDFs, and dioxin-like PCBs in human and other tissues. EPA missed the opportunity to quantify the effect of these differences in an uncertainty analysis of the current exposure estimates.

Exposures in Highly Exposed Populations or at Key Developmental Stages

As noted above, it is not clear to the committee why EPA relied entirely on a model with all its inherent uncertainties to conclude that the annual infant intake of TCDD, other dioxins, and DLCs from nursing significantly exceeds the currently estimated adult intake of 1 pg TEQ/kg/day. EPA failed to provide any measurements or environmental samples to support the conclusions drawn from the model. Providing this information would increase the confidence in its conclusions on this issue.

Major Assumptions

With regard to sources and emissions, the most appropriate way to characterize historical sources of TCDD, other dioxins, and DLCs is to compile a list of all known sources, make emissions estimates for each class from the available literature, and then combine these emissions to establish historical trends. The committee finds this assumption reasonable and sufficiently documented but finds that it would be valuable for EPA to consider alternative approaches (e.g., the top-down approach) for confirming or revising this approach.

In its consideration of highly exposed subpopulations, EPA found information indicating that breast-feeding might result in higher TCDD-TEQ body burdens of the nursing infant compared with those of non-nursing infants. The issue that exposure of the developing infant is already starting during pregnancy (in utero exposure) is not addressed in this section or not clearly mentioned in the full Reassessment. EPA did not consider this information in their overall conclusions about exposure. Moreover, because of the potential for causing anxiety among nursing mothers, EPA should expand its discussion about the multiple known benefits of breast-feeding as a footnote to the section describing exposures to nursing infants.

Modeling Assumptions

In characterizing exposures, EPA relied primarily on measurements combined with assumptions for emissions and relied almost completely on measurements of environmental and tissue levels for estimating exposure and body burdens. With the exception of their toxicokinetic model for nursing mothers, they did not rely on models for assessing transport and distribution from sources to environmental (such as air, water, and soil) and exposure (food products) media.

EPA's finding that, for purposes of environmental fate modeling, it is important to use the individual PCDD, PCDF, and PCB congener values rather than TEQ is self-evident and robust. The committee concurs with the EPA finding that TEQ should not be used in place of individual congener concentration as the variable in fate models for TCDD, other dioxins, and DLCs.

Were the Most Appropriate Studies Relied Upon?

For characterizing emissions, EPA developed a comprehensive inventory of all known emissions of TCDD, other dioxins, and DLCs but did not fully characterize the work of those researchers who looked at a top-down approach for characterizing historical emissions of PCDD and PCDF compounds. Rappe (1991), Harrad and Jones (1992), Brzuzy and Hites (1995), and Eisenberg et al. (1998) used fate modeling and mass-balance analyses applied to environmental samples and a number of assumptions to determine the extent to which deposition matches emissions.

CONCLUSIONS AND RECOMMENDATIONS

• To assess the total magnitude of emissions of TCDD, other dioxins, and DLCs, EPA used a bottom-up approach in which they attempt to identify all source categories and estimate the magnitude of emissions for that category. EPA also should use a top-down approach that attempts to account for observed levels and consider what emissions would be required to account for these levels. These alternative approaches give rise to significantly different estimates of the historical levels of emissions. Both approaches come with uncertainties. Thus, the readers of the Reassessment could benefit substantially from EPA using both approaches simultaneously to set plausible bounds on the historical and current trends in emissions.

- EPA needs to be explicit about how they dealt with measurements below the LOD in environmental and exposure media samples. Whether the less-than-LOD samples are assumed to be zero, assumed to be one-half LOD, imputed by fitting a censored regression model, or dealt with by using some other assumption could have significant impacts on estimates of TCDD, other dioxins, and DLC intakes. EPA should explicitly address how its assumption affects the magnitude and range of estimated intakes relative to alternative approaches. Moreover, EPA should describe how the changing LOD affects its estimate of the time trend of TEQ intake.
- Because many users of the Reassessment will be interested in reducing exposures to TCDD TEQ and identifying strategies for achieving reductions in body burden, EPA should add some discussion in the exposure chapter about what factors (such as diet, activities, and location) tend to increase or decrease TEQ intake.
- EPA should construct their reports so that information in the summary emissions inventory table of Part III can be more clearly and more easily traced back to the source chapters that provide background information.
- EPA should evaluate the impact on early emission-inventory estimates (1987, 1995) of sources added in more recent assessments (2005) so that the overall percentage declines reflect all sources. Such an evaluation would help to confirm dramatic decreases in TEQs that appear to have occurred over time.
- EPA should define a strategy for collection of samples and reanalysis of archived samples to answer a number of remaining questions about exposure trends and to fill in some important data gaps. (The committee does not consider it particularly useful or cost-effective for EPA to obtain and analyze more environmental media samples for the full range of TCDD, other dioxins, and DLCs.)
- EPA should create a congener-specific and active database of *typical* concentrations for the whole range of PCDDs, PCDFs, and dioxin-like PCBs (included in the WHO TEF list). This recommendation applies to work separate from the Reassessment. The database should be based on a compendium of all available data and be updated on a regular basis with new data as they are published in the peer-reviewed literature. Maintaining the database would not require EPA to conduct its own sampling program. Such a database would need to fulfill clear requirements of data quality and traceability, including chemical analysis, representative and targeted sampling, data representative of consumer exposure, presentation of data, handling, and presentation of less-than-LOD samples.
- In view of the number of sites with increased levels of PCBs in the

environment and anticipating that those levels could result in higher contributions of the dioxin-like PCB fraction to total TEQ exposure (e.g., through local fish consumption), EPA should explicitly characterize the variability of population exposures to PCBs. EPA should estimate the magnitude of the ratio of high-end to median and mean exposure, the factors (e.g., proximity to sources, geographic region, and eating habits) that give rise to high-end exposure, and the relative uncertainty with which high-end exposures can be estimated.

5

Cancer

This chapter reviews the U.S. Environmental Protection Agency (EPA) assessment of the carcinogenicity of 2,3,7,8-tetrachlorodibenzo-*p*-dioxin (TCDD), commonly referred to as dioxin, other dioxins, and dioxin-like compounds (DLCs), including EPA's qualitative characterization of their carcinogenicity, the assumption that the dose-response relationship is linear, and the use of animal bioassay and epidemiological data to quantify the dose response. The final section summarizes the committee's conclusions.[1]

QUALITATIVE EVALUATION OF CARCINOGENICITY

EPA concludes that dioxin is "carcinogenic to humans" based on the following evidence (Reassessment, Part III, pp. 6-7 to 6-8): evidence from the occupational cohort studies that dioxin exposure increases mortality from cancer aggregated over all sites and from lung cancer "and, perhaps, other sites"; evidence from bioassays of cancer in both sexes of multiple species at multiple sites; and evidence regarding dioxin's mode of action, including mechanistic evidence that dioxin acts as a tumor promoter via receptor-mediated pathway(s) and the finding that the receptor-mediated pathways that may give rise to cancer in laboratory animals appear to be present and functional in human tissues.

---

[1]*The Exposure and Human Health Reassessment of 2,3,7,8-Tetrachlorodibenzo-p-Dioxin (TCDD) and Related Compounds* (EPA 2003a, Part I; 2003b, Part II; 2003c, Part III) is collectively referred to as the Reassessment.

In this chapter, the committee reviews the epidemiological, bioassay, and mode of action evidence and then presents conclusions regarding both qualitative and quantitative measures of carcinogenicity of TCDD, other dioxins, and DLCs.

Epidemiological Evidence

The epidemiological evidence that provided the basis for EPA's assessment consists primarily of studies following four cohorts. Of these, the Reassessment reviewed in detail those related to the three cohorts that provided quantitative dose-response estimates linking serum dioxin to cancer mortality (Ott and Zober 1996; Becher et al. 1998; Steenland et al. 2001). The cohorts were quite variable in size and exposure ranges. Ott and Zober (1996) studied a relatively small number of men exposed to an accidental release of dioxin in 1953 (N = 243, 13 cancer deaths). Becher et al. (1998) examined a cohort of 1,189 men employed in pesticide and herbicide production, from which 124 cancer deaths were identified. The third cohort represents a large occupational population originally studied by Fingerhut et al. (1990, 1991), who examined 5,172 male employees in 12 manufacturing facilities. An update on this cohort was provided by Steenland et al. (1999), who applied "job-exposure matrix"[2] estimates to 5,132 workers in the original cohort who were followed for 6 more years. The total number of cancer deaths in this cohort was 377. In 2001, Steenland et al. updated this study again on a subcohort of 3,538 workers (with 256 cancer deaths) and used data from 170 members of this cohort for which estimated external exposures and known serum dioxin levels were available to establish a quantitative dose-response assessment.

Each study identified a cohort of workers who had been employed in industrial settings in which dioxin was a by-product. These settings included pesticide production (Ott and Zober 1996; Becher et al. 1998) or chemical plants more broadly (Steenland et al. 2001). In each instance, current serum dioxin measurements were available for a subset of workers. Development of exposure estimates for the entire cohort required two extrapolations: from current serum dioxin measurements to historical exposure levels using estimates of serum dioxin half-life, and from workers with current serum dioxin measurements to those without by linking available serum dioxin measurements to job characteristics based on knowledge of the industrial processes. Although these extrapolations decrease the accu-

[2]A "job-exposure matrix" refers to an algorithm by which experience in particular jobs are assigned estimated exposure levels. Each job (a row of the matrix) has a corresponding series of exposure levels assigned (columns).

racy of the assessment, they were necessary to provide historical exposure estimates so that there would be a sufficient number of cohort members who could be included in the analysis.

In addition to these three cohorts, Part II, section 7.5.4 of the Reassessment describes studies that reported on an occupational cohort of 2,310 workers in two plants that prepared and manufactured phenoxy herbicides in the Netherlands (see Reassessment, Part II, Table 7-21 for a summary of all four studies). Bueno de Mesquita et al. (1993) found no statistically significant increases in cancer mortality among all workers (31 deaths, standardized mortality ratio [SMR] = 107, 95% confidence interval [CI] = 73 to 152) and among a subset of 139 workers involved in a 1963 industrial accident (10 deaths, SMR = 137, 95% CI = 66 to 252). Comparing exposed workers (N = 963) to unexposed workers (N = 1,111), both total cancer mortality (rate ratio, [RR] = 1.7, 95% CI = 0.9 to 3.4) and respiratory cancer mortality (RR = 1.7, 95% CI = 0.5 to 6.3) were nonsignificantly increased.

A follow-up study by Hooiveld et al. (1996) reported a statistically significant increase in cancer mortality among workers in one of the two plants (SMR = 146, 95% CI = 109 to 192). No such increase was observed in the other factory. Follow-up analysis by Hooiveld et al. (1998) reported a statistically increased incidence of malignant neoplasms among 140 workers involved in the 1963 industrial accident (SMR = 1.7, 95% CI = 1.1 to 2.7). The incidence of malignant neoplasms was also increased in a larger group of 549 workers (SMR = 1.5, 95% CI = 1.1 to 1.9). A comparison of this group of 549 exposed workers to 482 unexposed workers, also from this cohort, yielded an increased total cancer mortality risk (RR = 4.1, 95% CI = 1.8 to 9.0) and an increased respiratory cancer mortality risk (SMR = 7.5, 95% CI = 1.0 to 56.1).

There are three major issues to consider regarding EPA's review of the epidemiological studies investigating the relationship between dioxin exposure and cancer. First, although EPA identified the cohort studies capable of generating quantitative dose-response information for the dose-response modeling and considered the broader epidemiological literature in the background documents, Part III of the Reassessment did not provide a thorough and systematic analysis of the body of epidemiological evidence from which these three studies were chosen. In particular, although Part II described the complete array of studies, including those by Kogevinas et al. (1997) and Bertazzi et al. (1998), the Reassessment did not analyze site-specific tumors consistently across all studies but rather emphasized the positive findings in each paper without a full discussion of consistency, or lack thereof, across studies.

A second issue is EPA's decision to focus on total cancers instead of specific types of cancer. EPA argues that because dioxin is not genotoxic and is instead presumed to act primarily as a promoter rather than an initiator of cancer, the lack of specificity in tumor type is to be expected. If dioxin promotes cancer through the Ah receptor mechanism, however, then an increased tumor incidence would require expression of the receptor in that tissue. The Ah receptor is expressed in most tissues but to varying degrees. It is uncertain whether the level of expression is an important determinant of tumor promotion. There are also many downstream events from ligand-receptor interaction that are tissue specific and essential for tumor promotion to occur via a receptor-mediated response, and these downstream events differ from tissue to tissue (see also discussion on mode of action later in this chapter). In any case, EPA reasons that, in the face of limited power, increased risk of total cancers (which would reflect the increased incidence across the multiple sites affected by dioxin) is easier to detect than an increased risk of individual cancer types (see Part III, pp. 2-9 to 2-10). This rationale would be valid for a given relative risk (e.g., a doubling of the incidence or mortality). However, a given absolute incremental risk (e.g., an additional 10 cancers due to exposure) would be more readily identified for a specific cancer site than for cancers in the aggregate.

The more compelling argument for aggregating across cancer types is the practical one that the results for specific cancers are extremely imprecise in these cohorts of modest size. If, in fact, multiple cancer types all showed a small increment in risk of equal magnitude, there would be greater precision and statistical power for the aggregation. For example, in the case of ionizing radiation, the aggregation across a series of radiosensitive cancers, each with small increases in risk, yields a more statistically precise indication of an increase in cancers related to radiation exposure than do any of the individual cancers. In the case of dioxin, it is not clear that a specific set of cancers is affected that can then be aggregated to enhance statistical power.

To evaluate the patterns across cancer sites, the committee examined selected papers from the three cohorts (Ott and Zober 1996; Flesch-Janys et al. 1998; Steenland et al. 1999). This evaluation revealed that only limited information is available regarding numbers of cases at specific sites, hence limiting the opportunity to examine consistency across studies. As noted by others, there is some consistency across studies for respiratory cancers, but there is a general lack of concordance for the other cancer sites reported in more than one study. The degree of replication or lack thereof should not be overstated given the small number of studies and imprecise information on specific cancer sites from all but the Steenland et al. (1999) report.

Overall, the committee concurs with the value of conducting analyses of total cancers, given the potential for dioxin to affect multiple types of cancer and the limited precision of risk estimates for individual cancer types. Nonetheless, the potential for effects limited to specific types of cancer, as has been found for other causes, also warrants an analysis of major cancer types (e.g., respiratory cancers), the imprecision notwithstanding.

Another concern is the potential role of confounding by lifestyle factors such as smoking and by occupational exposures that co-occur with dioxin. Although smoking is a powerful lung carcinogen, quite capable of generating spurious relative risks on the order of those reported in the epidemiological studies for dioxin of around 1.5, the design of those studies makes its potential role as a confounder unlikely in this case. The key comparisons were not between industrial workers and the general population, which is quite susceptible to confounding by lifestyle factors, but among subsets of workers with different levels of estimated dioxin exposure. It is not likely that smoking histories would differ markedly among men located at different jobs *within* the industrial plant or in relation to duration of employment. In contrast, there is greater potential for confounding by other workplace agents given that the industrial cohorts had exposure to pesticides and potentially carcinogenic chemicals in addition to dioxin. Although these accompanying workplace hazards likely differed for the three cohorts that contributed to the quantitative risk assessment, confounding could have occurred in each to yield a similar falsely elevated measure of association. The difficulties in isolating the health effects of single agents from the complex mixtures encountered in chemical manufacturing must be recognized.

Epidemiological evidence for an association between cancer and exposure to DLCs has been characterized as "inadequate but suggestive" (EPA 1987) and "limited" (IARC 1997). ATSDR (2000) concluded that the epidemiological evidence "taken in totality, indicates a potential cancer causing effect for PCBs."

On the whole, it was the committee's impression that EPA's narrative in discussing epidemiological studies in Part III of the Reassessment tended to focus on positive findings without fully considering the strengths and limitations of both positive and negative findings. Part III of the Reassessment would be strengthened if EPA clearly identified specific inclusion criteria for those studies for which quantitative risk estimates were determined.

Bioassay Data

Several large and well-conducted dioxin-related cancer bioassays (Kociba et al. 1978; NTP 1982a,b; NTP 2004) have reported induction of several types of cancer in both rats and mice. The study in hamsters was confounded by use of dioxane, which is a potential carcinogen, as the delivery vehicle. Table 5-1 summarizes these studies. In all studies in which dioxin elicited an increase in tumors, the increase was site specific. With oral administration, the organ most frequently affected was the liver, reflecting the mode of action of carcinogenicity, as discussed below.

Of the 21 DLCs of concern, EPA (Part II, p. 6-30) reported that carcinogenicity bioassays have been conducted on only two pure polychlorinated dibenzo-*p*-dioxin (PCDD) and 1,2,3,7,8-pentachlorodibenzo-*p*-dioxin (PeCDD), and a mixture of two congeners (1,2,3,6,7,8- and 1,2,3,7,8,9-hexachlorodibenzo-*p*-dioxin [HxCDD]). Carcinogenicity bioassays have also been conducted on one polychlorinated dibenzofuran (PCDF) (2,3,4,7,8-pentachlorodibenzofuran [PeCDF]) and one PCB (126; 3,3′,4,4′, 5-pentachlorobiphenyl (PeCB) (Table 5-2).

However, the ability of a variety of dioxins other than TCDD and DLCs to enhance the carcinogenicity of known carcinogens (promoter assays) has also been reported for PeCDD, HpCDD, 2,3,7,8-tetrachlorodibenzofuran (TeCDF), PeCDF, and 1,2,3,4,7,8-hexachlorodibenzofuran (HxCDF) (summarized by IARC 1997). Bioassays have also been conducted on mixtures of PCBs, and although they provide some information on the carcinogenicity of components, they do not identify the specific responsible chemical(s).

Mode of Action

Dioxin does not have structural features that would lead to a reactive electrophile, and it is clearly not DNA reactive, as no DNA binding or adducts were found in rodent tissues (Poland and Glover 1979; Randerath et al. 1988; Turteltaub et al. 1990). Absence of DNA reactivity is supported by negative findings in genetic toxicological assays (IARC 1997).

Nevertheless, EPA notes (Part II, p. 6-1) the hypothesis that dioxin might be indirectly genotoxic, either through induction of oxidative stress or by altering the DNA damaging potential of some endogenous compounds, including estrogens. No evidence is available for estrogen-mediated DNA damage resulting from dioxin exposure, but oxidative DNA damage has been documented after 30 weeks administration of dioxin (Tritscher et al. 1996; Wyde et al. 2001). Indirect genotoxicity has been postulated to initiate carcinogenicity, but there is insufficient evidence that dioxin has initiating activity.

Dioxin was reported to have weak initiating activity in one study (DiGiovanni et al. 1977) in which it was applied to mouse skin prior to a promoting agent. This finding has not been corroborated, and in contrast to what would be expected from an initiating agent, application of dioxin to mouse skin at a dosage greater than that required for a promoting effect did not induce skin tumors (Poland et al. 1982).

Moreover, dioxin has not been specifically tested as an initiator in standard models in rat or mouse liver in which chemicals can be evaluated

TABLE 5-1 Dioxin Cancer Bioassays

Species/Strain	Route and Dose	Sex	Sites of Tumor Increases	Reference
Rat/Sprague-Dawley	Oral in feed 1, 10, 100 μg/kg/day	Male	Oral cavity	Kociba et al. 1978
		Female	Lung, oral cavity, liver	NTP 1982a
Rat/Osborne Mendel	Gastric instillation 10, 50, 500 μg/kg/week for 104 weeks	Male	Thyroid	
		Female	Liver	
Rat/Sprague-Dawley	Gastric instillation 3, 10, 22, 46, or 100 mg/kg, 5 days/week for 104 weeks	Female	Liver, lung, oral cavity, uterus	NTP 2005
Mouse/B6C3F1	Gastric instillation 0.01, 0.05, 0.5 mg/kg/wk for 104 weeks (males)	Male	Liver	NTP 1982a
	0.04, 0.2, 2.0 mg/kg/wk for 104 weeks (females)	Female	Liver, thyroid	
Mouse/Swiss Webster	Topical application 0.005 μg 3 days/week for 104 weeks	Female	Skin	NTP 1982b

Mouse/B6C3 and B6C	Intraperitoneal injection 1, 30, 60 μg/kg/week for 5 weeks	Male	Thymus (both), liver (B6C3 only)	DellaPorta et al. 1987
		Female	Thymus (both), liver (B6C3)	
Mouse/B6C3	Gastric instillation 2.5, 5.9 μg/kg/week for 52 weeks	Male	Liver	DellaPorta et al. 1987
		Female	Liver	
Mouse/Swiss	Gastric instillation 0.007, 0.7, 7.0 μg/kg/week for 52 weeks	Male	Liver	Toth et al. 1979
Mouse/TG.AC	Topical application for 24 weeks	Male	Skin papillomas	Eastin et al. 1998
		Female	Skin papillomas	
Mouse/TP53$^{+/-}$	Gastric instillation 250 μg/kg, 1,000 μg/kg twice weekly for 24 weeks	Male	None	Eastin et al. 1998
		Female	None	
Hamster/Syrian Golden	Intraperitoneal or subcutaneous injection 50 or 100 μg/kg every 4 weeks	Male	Skin	Rao et al. 1988

TABLE 5-2 TCDD, Other Dioxins, and DLC Cancer Bioassays

Congener	Bioassay
Dioxins	
2,3,7,8-TCDD	Rat (M,F)/mouse (M,F)
1,2,3,7,8-PeCDD	Rat (M,F)/mouse (M,F)/promoter rat (F)
1,2,3,4,7,8-HxCDD	No bioassay conducted
1,2,3,6,7,8-HxCDD	Combination study
1,2,3,7,8,9-HxCDD	Combination study
1,2,3,4,6,7,8-HpCDD	Promoter rat (F)
1,2,3,6,7,8- and 1,2,3,7,8,9-HxCDD mix	Rat (M,F)/mouse (M,F)
OCDD	No bioassay conducted
Furans	
2,3,7,8-TCDF	Promoter mouse (F)
1,2,3,7,8-PeCDF	No bioassay conducted
2,3,4,7,8-PeCDF	Rat (F)/promoter mouse (F) and rat (M)
1,2,3,4,7,8-HxCDF	Promoter mouse (F)/rat (M)
1,2,3,6,7,8-HxCDF	No bioassay conducted
1,2,3,7,8,9-HxCDF	No bioassay conducted
2,3,4,6,7,8-HxCDF	No bioassay conducted
1,2,3,4,6,7,8-HpCDF	No bioassay conducted
1,2,3,4,7,8,9-HpCDF	No bioassay conducted
OCDF	No bioassay conducted
Non-ortho PCBs	
3,3′,4,4′-TCB (77)[a]	No bioassay conducted
3,4,4′,5-TCB (81)	No bioassay conducted
3,3′,4,4′,5-PeCB (126)	Rat (F)
3,3′,4,4′,5,5′-HxCB (169)	No Bioassay Conducted
Mono-ortho PCBs	
2,3,3′,4,4′-PeCB (105)	No Bioassay Conducted
2,3,4,4′,5-PeCB (114)	No Bioassay Conducted
2,3′,4,4′,5-PeCB (118)	No Bioassay Conducted
2′,3,4,4′,5-PeCB (123)	No Bioassay Conducted
2,3,3′,4,4′,5-HxCB (156)	No Bioassay Conducted
2,3,3′,4,4′,5′-HxCB (157)	No Bioassay Conducted
2,3′,4,4′,5,5′-HxCB (167)	No Bioassay Conducted
2,3,3′,4,4′,5,5′-HpCB (189)	No Bioassay Conducted

[a]International Union of Pure and Applied Chemistry numbers in parentheses.
Abbreviations: OCDD, octachlorodibenzo-*p*-dioxin; OCDF, octachlorodibenzofuran; TCB, -tetrachlorobiphenyl.

as initiators followed by administration of a promoting substance (Enzmann et al. 1998). Also, in several chronic bioassay studies in which dioxin was administered to female Sprague-Dawley rats for 30 weeks at dosages associated with an increased incidence of liver tumors in carcinogenicity studies, no increase in hepatic preneoplastic lesions indicative of initiation was found (Lucier et al. 1991; Maronpot et al. 1993). Thus, at present, there is no direct experimental evidence that dioxin acts as an initiator in rat liver.

A lack of initiating activity would be consistent with an absence of direct genotoxicity (Williams 1992). Nevertheless, some dose-response modeling of data that show a promoting effect of dioxin on rat liver preneoplastic lesions suggested that dioxin also had "a weak" (Moolgavkar and Luebeck, 1995) or "a slight" (Portier et al. 1996) initiating effect. In contrast, analysis of a two-cell clonal growth model reproduced such data without presuming an effect on mutation rates (that is, initiation) (Conolly and Andersen 1997).

Resolution of the question of initiating activity of dioxin awaits experimental evidence. Also, the postulated linkage between potential initiating activity and oxidative DNA damage is not established. In an investigation of the mode of action of hepatocarcinogenicity of pentachlorophenol, oxidative DNA damage was not found to produce liver initiation (Umemura et al. 1999).

The committee agrees with EPA that TCDD, other dioxins, and DLCs appear to enhance tumor development in female rat liver via tumor promotion. The promoting activity and liver tumor-enhancing activity of dioxin seem to be mediated through activation of the Ah receptor (aromatic hydrocarbon receptor [AHR]), which in turn leads to a variety of changes in gene expression, including notably induction of cytochromes P450 (CYPs) (Whitlock 1989) and genes related to cell proliferation (Puga et al. 1992) (see Figure 5-1). Whether those gene changes mediate the reported oxidative stress is not known. Nevertheless, both CYP induction and oxidative stress could be involved in liver cytotoxicity, which was found in studies that examined this parameter (Maronpot et al. 1993; Viluksela et al. 2000). Cytotoxicity, in turn, elicits regenerative cell proliferation (Williams and Iatropoulos 2002), as reported in several dioxin studies (Lucier et al. 1991). Dioxin-induced changes in gene expression, however, can occur without enhancement of hepatocelluar proliferation (Fox et al. 1993). In fact, increases in cell proliferation have been documented only after 30 weeks of dioxin administration (Lucier et al. 1991). The enhanced cell proliferation arising from either altered gene expression or cytotoxicity or both could be the principal factor leading to promotion of hepatocellular tumors (Busser and Lutz 1987; Whysner and Williams 1996). The sensitivity of female rat liver to dioxin, which apparently does not extend to the mouse, clearly depends on ovarian hormones (Lucier et al. 1991; Wyde et al. 2001). This sensitivity has been ascribed to induction of estradiol metabolizing enzymes (Graham et al. 1988) and is hypothesized to lead either to generation of reactive metabolites of endogenous estrogen or to active oxygen species of estrogens. Oxidative DNA damage has been implicated in liver tumor promotion (Umemura et al. 1999). In contrast to the extensive work on hepa-

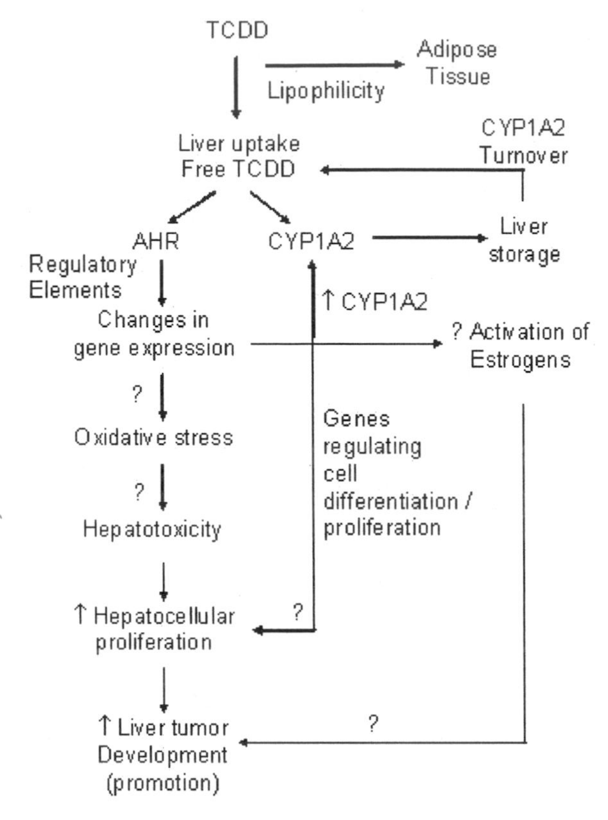

FIGURE 5-1 Possible mechanism for TCDD hepatocarcinogenicity.

tocellular neoplasia, little is known about the pathogenesis of the bile-duct tumors (Table 5-3).

Mechanistic issues are discussed in greater detail below in the context of evaluating whether the dose-response relationship is likely to be linear. In any case, the committee agrees with EPA's general conclusion that there is sufficient evidence from epidemiological studies, animal bioassays, and mode of action studies to support the qualitative conclusion that TCDD, other dioxins, and DLCs are likely to cause cancer in humans with adequate conditions of dose and duration of exposure.

Committee's Perspective on Whether the Scientific Evidence Supports Classification of Dioxin As a Known Human Carcinogen

After extensive discussion of EPA's revised definition of "carcinogenic to humans" and "likely to be carcinogenic to humans" provided in EPA's

TABLE 5-3 Dioxin Rat Bioassays

Study End Point	Kociba et al. (1978) (1, 10, 100 ng/kg/day)	NTP (1982a) (3, 10, 100 ng/kg/wk)	NTP (2005) (10, 22, 46, 100 ng/kg)
Survival	Decreased at 100 ng/kg	No effect	
NOEL for tumors	1 ng/kg		10.22 ng/kg
Liver	Adenoma/carcinoma (58%); bile-duct adenoma, some in low dose	Cholangiocarcinoma, adenoma (30%), cholangioma, hepatocholangioma	Cholangiocarcinoma (46, 100 ng/kg), adenoma (100 ng/kg)
Lung	Keratinizing squamous cell carcinoma (100 ng/kb)	Cystic keratinizing epithelioma (100 ng/kg)	Cystic keratinizing epithelioma (100 ng/ kg)
Oral cavity	Squamous cell carcinoma, hard palate	Squamous cell carcinoma, gingival	Squamous cell hyperplasia (all)

Abbreviation: NOEL, no-observed-effect level.

2005 *Guidelines for Carcinogen Risk Assessment* (EPA 2005a, see also Appendix B) and consideration of the points above, the committee agrees that there is strong and convincing evidence that dioxin is likely to be a human carcinogen. Although the committee does not reject outright the somewhat higher classification of dioxin as "carcinogenic to humans," there was not unanimous agreement that the available scientific information on human dioxin carcinogenicity met condition (a) of EPA's cancer guidelines, which states "there is strong evidence of an association between human exposure and either cancer or the key precursor events of the agent's mode of action but not enough for a causal association" (EPA 2005a, p. 2-54).

The committee was in general agreement that the epidemiological evidence, although not "strong," was generally consistent with a positive association between occupational dioxin exposure and mortality from all cancers, but the magnitude of the effect was modest, and the limited evidence for any specific tumor type being significantly associated was of some concern. This conclusion is in fact quite similar to EPA's assessment of the relative strength of the epidemiological evidence (Reassessment, Part III, p. 2-21). In its discussion, the committee remained uncertain about the intent of the language in the 2005 *Guidelines for Carcinogen Risk Assessment* stating that condition (a) could be satisfied if there is "strong evidence of an association between human exposure and either cancer or the key precursor events of the agent's mode of action but not enough for a causal association" (EPA 2005a, p. 2-54). The committee agreed that there is convincing evidence supporting the interaction of dioxin with the human Ah receptor and that the interaction with the receptor was necessary, but not sufficient, to cause cancer in animals. However, the committee was not in complete agreement about whether these conditions met the stated criterion of a "key precursor event of the agent's mode of action" (EPA 2005a, p. 2-54). For example, it was noted that, even though TCDD binds to the human Ah receptor, several endogenous and exogeous substances, including bilirubin, biliverdin, and β-naphthoflavone, also bind to the Ah receptor but are not carcinogenic in rodent models (Seidel et al. 2000); hence, some other key precursor event(s) may need to be identified to meet that criterion. However, it was also recognized that persistence of the Ah receptor activation may be a key determinant for carcinogenicity because genetic modification of the AhR gene, causing activation in the absence of any ligand, results in a tumorigenic response in mice (Andersson et al. 2002).

Furthermore, there is evidence that prolonged stimulation of AHR by the nonpersistent ligand indole-3-carbinol (or its acid condensation products; derived from broccoli and other cruciferous vegetables) can promote a variety of tumor types after initiation with different genotoxic compounds (Pence et al. 1986; Bailey et al. 1987; Dashwood et al. 1991; Kim et al.

1997; Dashwood 1998; Yoshida et al. 2004). The committee recommends that EPA use the 2005 *Guidelines for Carcinogen Risk Assessment* (EPA 2005a, see also Appendix B) specifically in its final assessment and carefully delineate its interpretation of what constitutes "strong" evidence and a "key precursor event" under condition (a) of the definition of "carcinogenic to humans."

The committee noted that classification artificially places an apparent bright line distinguishing a substance as "carcinogenic to humans" and "likely to be carcinogenic to humans," whereas the actual scientific evidence lies on a continuum. In the context of a weight-of-evidence continuum, the committee found that the scientific evidence favored the high end of the "likely to be a human carcinogen" classification or the lower end of the "carcinogenic to humans" classification and emphasized that dioxin remains unique with respect to the International Agency for Research on Cancer (IARC) Group 1 designation based to a large extent on total cancers instead of a specific cancer type.

The committee recognizes that the 2003 Reassessment used a different definition of "carcinogenic to humans" based on EPA's 2003 draft carcinogen risk assessment guidelines. The committee found that the argument provided by EPA in the 2003 Reassessment to support its position that the epidemiological data met the criterion of "strong evidence of an association" between dioxin exposure and cancer risk was unconvincing. However, the committee questioned whether it is worth EPA's investment of significant efforts to further qualitatively classify the carcinogenicity of dioxin. The committee considers that quantitative risk estimates for dioxin should not depend on which side of the artificial bright line between "likely to be a human carcinogen" and "carcinogenic to humans" EPA ultimately places dioxin. The committee urges EPA to focus on improved quantitative characterization of risks and to reduce the emphasis on qualitative characterization of hazard in this case.

QUANTITATIVE CONSIDERATIONS IN ASSESSING TCDD, OTHER DIOXINS, AND DLC CARCINOGENICITY

EPA's Assumption That the Dose-Response Relationship Is Linear

To estimate a cancer slope factor (CSF) for dioxin using either animal bioassay data or epidemiological data, EPA estimated a point of departure (POD) dose as the dose yielding an excess cancer risk of 1% and then extrapolated back to zero incremental dose using a straight line. The dose (mg/kg-day) corresponding to a 1% excess cancer risk is referred to as the ED_{01} (effective dose). Dividing the ED_{01} dose into 0.01 yields the CSF, expressed in units of $(mg/kg\text{-}day)^{-1}$. A more conservative, but widely used, estimate of the CSF can be calculated by estimating the lower confidence limit on the ED_{01} (designated the LED_{01} [lower confidence bound on the effective dose]) and then dividing that value into 0.01.

EPA describes its approach as follows (Part III, p. 5-15):

> Extrapolation from the POD to lower doses is conducted using a straight line drawn from the POD to the origin—zero incremental dose, zero incremental response—to give a probability of extra risk. The linear default is selected on the basis of the agent's mode of action when the linear model cannot be rejected and there is insufficient evidence to support an assumption of non-linearity.

Because EPA's assumption of linearity at doses below the 1% excess risk level for carcinogenic effects of TCDD, other dioxins, and DLCs is central to the ultimate determination of regulatory values, it is important to critically address the available scientific evidence on the most plausible shape of the dose-response relationship at doses below the POD (LED_{01}). On the basis of a review of the literature, including the detailed review prepared by EPA and presented in Part II of EPA's Dioxin Risk Assessment and new literature available since the last EPA review, the committee concludes that, although it is not possible to scientifically prove the absence of linearity at low doses, the scientific evidence, based largely on mode of action, is adequate to favor the use of a nonlinear model that would include a threshold response over the use of the default linear assumption. The committee concludes that four major considerations of the scientific evidence support the use of a nonlinear model for low-dose extrapolation.

TCDD, Other Dioxins, and DLCs Are Not Directly Genotoxic

As noted earlier, available evidence suggests that TCDD, other dioxins, and DLCs are not directly genotoxic. There is general consensus in the scientific community that nongenotoxic carcinogens that act as tumor promoters exhibit nonlinear dose-response relationships, and that thresholds (doses below which the expected response would be zero) are likely to be present. In addition, even among compounds that covalently react with DNA, the dose response may be nonlinear (Williams et al. 2005). For example, the ED01 study (Staffa and Mehlman 1979) used more than 24,000 mice to evaluate the shape of the dose-response relationship over a 5-fold range of administered dose (30 to 150 parts per million) of the potent carcinogen 2-acetylaminofluorene, which is metabolized to a highly genotoxic metabolite that forms DNA adducts. The results of the lifetime feeding study showed a dose-related increase in bladder and liver tumors. The dose-response relationship for the liver tumors appeared to be linear, whereas the bladder tumor dose-response was markedly sublinear at the lower end of the curve.

Receptor-Mediated Agents Have Sublinear Dose-Response Relationships

Some studies suggest that TCDD (and, presumably, other dioxins and DLCs) may cause DNA damage indirectly via generation of reactive oxygen species that may result in oxidative DNA damage and intrachromosomal recombination, although, as noted above, initiating activity has not been demonstrated. However, these effects are secondary to a series of downstream events that are secondary to Ah receptor activation, a phenomenon that would be likely to cause the dose-response relationship to be sublinear at low doses. It is recognized that a roughly linear increase in response with increasing dose will occur at doses above a minimal response level (e.g., 1% or 5% excess risk), as would be expected for any receptor-mediated response. These comments are focused on extrapolation of the dose-response relationship to doses well below those associated with a minimum response level (POD).

The observation that adverse effects caused by TCDD, other dioxins, and DLCs depend on AHR activation underlies mechanistic considerations for these compounds. Part III, p. 3-1 (lines 5 to 14), of the Reassessment states that

> much evidence indicates that TCDD acts via an intracellular protein (the AhR) which functions as a ligand-dependent transcription factor in partnership with a second protein (ARNT [AHR nuclear translocator protein]). Therefore, from a mechanistic standpoint, TCDD's adverse effects appear likely to reflect alterations in gene expression that occur at an inappropriate time and/or for an inappropriate long time. Mechanistic studies also indicate that several other proteins contribute to TCDD's gene regulatory effects and that the response to TCDD probably involves a relatively complex interplay between multiple genetic and environmental factors. If TCDD operates through such a mechanism, as all evidence indicates, then there are certain constraints on the possible models than can plausibly account for TCDD's biological effects, and, therefore, on the assumptions used during the risk assessment process.

EPA cites further mechanistic studies describing interactions between the AHR and other critical regulatory proteins and transcription factors (Rb, SIM, HIF1-α, REL-A, among others) as evidence of the complex interplay between dioxin and other genetic and environmental factors. Table 3-1 in the Reassessment appropriately describes early molecular events.

There is widespread agreement in the scientific community that all or nearly all the adverse effects of TCDD, other dioxins, and DLCs depend on a receptor-mediated mechanism. Both IARC and EPA (see above) conclude that these compounds act through a mechanism involving the AHR. As noted in the Reassessment (Part III, p. 2-19, lines 28 to 33):

> Despite this lack of a defined mechanism at the molecular level, there is a consensus that 2,3,7,8-TCDD and related compounds are receptor-mediated carcinogens in that (1) interaction with the AhR is a necessary early event, (2) 2,3,7,8-TCDD modifies a number of receptor and hormone systems involved in cell growth and differentiation, such as the EGFR [epidermal growth factor receptor] and estrogen receptor, (3) sex hormones exert a profound influence on the carcinogenic action of 2,3,7,8-TCDD.

Mechanistic considerations for DLCs and dioxins other than TCDD are less well established, although it is widely held that most of the toxic and carcinogenic effects of other dioxins and DLCs are mediated via the same receptor signaling pathways as those for TCDD. The Reassessment cites "comparative binding studies and other data" (Part III, p. 2-3, lines 25 and 26) to suggest that DLCs and dioxins other than TCDD exhibit TCDD-like responses in proportion to their receptor binding affinity (generally reflected in their toxic equivalency factors [TEFs]). Although this association may hold for most toxic and biochemical responses, there are few, if any, biochemical or mechanistic studies describing interactions when DLCs and dioxins other than TCDD are the ligands/inducers. It is not clear if those interactions play a role in the adverse health effects of TCDD, other dioxins, and DLCs, nor have such interactions been characterized.

There is a large body of scientific data on receptor-mediated responses. However, while the relationship of receptor binding and effects on tumor development in rodents remains uncertain from a mechanistic point of view (Whysner and Williams 1996), the recent National Toxicology Program (NTP) bioassay results using the TEF/TEQ (toxic equivalent quotient) approach (which is dictated by AHR binding) strongly supports the role of AHR in hepatocarcinogenicity of DLCs. Receptor binding appears necessary, but insufficient, because many tissues with receptors are not sites of TCDD-induced (or, by inference, induced by other TCDD, other dioxins, and DLCs) preneoplastic changes or tumors.

A fundamental concept in pharmacology is that receptor-mediated responses show sigmoidicity in the shape of the log dose-response relationship, although ligand-receptor interactions and subsequent "down stream" events that ultimately produce drug efficacy or toxicity are complex (Ross and Kenalkin 2001). Response is a function of the number of occupied and activated receptors, which typically exhibit steep dose-response relationships. For example, Kohn and Melnick (2002) modeled the shape of the dose-response relationship for receptor-mediated responses, using the estrogen receptor and various xenoestrogens as a model receptor and ligands, respectively. The model included a variety of assumptions with regard to receptor number, ligand binding affinity, and partial agonist activities, yet in every instance clear sublinear responses were observed at low doses. In all instances modeled, a predicted response indistinguishable from the background response was seen at doses less than one order of magnitude lower than the dose providing the lowest detectable response (conceptually similar to a POD). The model parameters were based on ligands with relatively short half-lives and reversible binding to the receptor and thus may not be directly applicable to TCDD, other dioxins, and DLC binding to AHR.

Carcinogenicity of DLCs is not solely and quantitatively related to receptor binding. There are numerous synthetic and naturally occurring AHR ligands, to which humans are exposed through diet and the environment, that bind to and activate the receptor (and induce a transcriptional response as measured by cytochrome P4501A protein [CYP1A] mRNA and enzyme activity) and yet do not seem to act as tumor promoters or directly produce AHR-dependent toxic responses commonly seen with TCDD, other dioxins, and DLCs. Thus, although binding to and activation of AHR appears to be required for tumor promotion, it is not sufficient. On the other hand, others (Carney et al. 2004) have shown that morpholinos to AHR block cardiovascular toxicity in zebrafish, but morpholinos to CYP1A do not. This conclusion strongly suggests that additional downstream events are critical to the promotional effects of these chemicals. Because multiple additional steps are necessary, each probably with homeostatic mechanisms functional at low doses but perhaps overwhelmed at high doses, sublinearity with a response approaching zero at low doses would be expected.

EPA determined in previous evaluations of receptor-mediated carcinogens that a nonlinear, low-dose model, that may accommodate a threshold is appropriate. For example, numerous pesticides found to cause thyroid cancer secondary to modulation of thyroid hormone levels have been evaluated as threshold-type carcinogens (EPA 1998). In the recent NTP studies with dioxin, the observed thyroid tumors are undoubtedly due to perturbation of thyroid homeostasis (NTP 2005). Similarly, the induction of liver tumors from peroxisome proliferators was also deemed to occur via a threshold-type response but was further deemed largely irrelevant to humans because of species differences in peroxisome proliferator activated receptor (PPAR) function (EPA 2003d).

The final cancer guidelines (EPA 2005a, see also Appendix B) provide the following guidance on choosing between linear and nonlinear risk extrapolation approaches: "A nonlinear approach should be selected when there are sufficient data to ascertain the mode of action and conclude that it is not linear at low doses and the agent does not demonstrate mutagenic or other activity consistent with linearity at low doses" (p. 3-22). This is an important decision, as it will influence the methodology adopted in subsequent risk assessments. The final EPA cancer guidelines also make the following statement about risk assessment for carcinogens with a nonlinear mode of action (EPA 2005a, p. 3-20).

TABLE 5-4 Hepatic Toxicity in TCDD Rat Bioassays

Kociba et al. (1978)	NTP (1982a)	NTP (2005)
0 ng/kg, severity = 0.6 (57%)	0 ng/kg, 0 incidence	0 ng/kg
1 ng/kg, severity = 1.2 (88%)		
	3 ng/kg, severity = 1.0 (4%)	
10 ng/kg, severity = 2.1 (95%)	10 ng/kg, severity = 1.3 (15%)	10 ng/kg severity + 22 ng/kg severity 2+46 ng/kg severity 3+
100 ng/kg, severity = 3.6 (100%)	100 ng/kg, severity = 3.5 (100%)	100 ng/kg severity 4+

> At this time, safety assessment is the default approach for tumors that arise through a nonlinear mode of action; however, EPA continues to explore methods for quantifying dose-response relationships over a range of environmental exposure levels for tumors that arise through a nonlinear mode of action. (EPA 2002)

Evidence That Liver Tumors Are Secondary to Hepatotoxicity

In the Reassessment, EPA used the female rat liver tumor data from the Kociba et al. (1978) study to develop a dose-response relationship. In that study, the liver was the main site of carcinogenic activity (see Table 5-3).

Dioxin is retained preferentially in the liver in rats (Fries and Marrow 1975; Kociba et al. 1978), in addition to adipose tissue, which may underlie the liver susceptibility. In the rat liver, hepatic toxicity was accompanied by increases in liver tumors (Table 5-4), and numerous studies have shown that hepatotoxicity results in increased cell proliferation (Williams and Iatropoulos 2002). In the most recent dioxin bioassay (NTP 2004), toxic hepatopathy was found at 31 weeks at 100 mg/kg and at 53 weeks at 46 mg/kg and 100 mg/kg, but not at low dosages. Hepatocellular labeling indices were consistently elevated at these dosages at 31 and 53 weeks. In other studies, hepatotoxicity was less pronounced in male rats, for which no increase in tumors was seen. The hepatocarcinogenicity in female rats is related to estrogens and may be due to elevation of estrogen catechol levels resulting from AHR-dependent induction of cytochromes P450 in the CYP1 family responsible for generating catechols from estradiol. Accordingly, toxicity and cell proliferation may have been key events for hepatocarcinogenicity in these studies, as has been delineated for a variety of other rodent hepatocarcinogens (Williams 1997).

The cancer guidelines (EPA 2005a, see also Appendix B) caution against using tumor data for quantitative, low-dose extrapolation when clear evidence of cytotoxicity is present:

> Studies that show tumor effects only at excessive doses may be compromised and may or may not carry weight, depending on the interpretation in the context of other study results and other lines of evidence. Results of such studies, however, are generally not considered suitable for dose-response extrapolation if it is determined that the mode(s) of action underlying the tumorigenic responses at high doses is not operative at lower doses. . . . Studies that show tumors at lower doses, even though the high dose is excessive and may be discounted, should be evaluated on their own merits. (EPA 2005a, p. 2-18)

Earlier in the document, EPA states,

> In addition, overt toxicity or altered toxicokinetics due to excessively high doses may result in tumor effects that are secondary to the toxicity rather than directly attributable to the agent. (EPA 2005a, p. 2-17)

Thus, based on these criteria, evidence of substantial hepatoxicity in tumor-bearing animals would raise questions about the use of hepatic tumors in female rats for quantitative, low dose extrapolation.

Although there is evidence that the liver tumors observed may be due to hepatotoxicity and have a sublinear dose-response relationship, the committee notes that two other types of epithelial tumors (keratinizing epithelioma of the lung and squamous cell tumors of the oral mucosal epithelium) were increased in a dose-dependent manner with no apparent indication of cytotoxicity in these tissues. However, the shape of the dose-response relationship for these tumors suggests that they may be nonlinear, as described below.

Bioassay Evidence of Nonlinearity

The recent NTP bioassay data (NTP 2004; Walker et al. 2005) show a consistent sigmoidicity to the tumor dose response. Walker et al. (2005) reported a Hill coefficient[3] of 2.81 (standard error [SE] = 0.68) for cholangiocarcinoma, 3.74 (SE = 1.5) for hepatocellular adenoma, 23.4 (SE insufficiently stable to report) for keratinizing epithelioma of the lung, and 2.14 (SE insufficiently stable to report) for squamous cell tumors of the oral mucosal epithelium. The central estimates for these coefficients all exceed

[3]The Hill function is defined as $f(dose) = dose^n/(k^n + dose^n)$ and can be used as a model component for dose effects. The coefficient $n > 1$ signifies a departure from linearity at lower doses.

2, hence suggesting nonlinearity, although the 95% CIs do not exclude a Hill coefficient of 1, which corresponds approximately to a linear dose response at low doses. Nonetheless, although the data alone do not rule out a linear tumor response at doses below a 5% response level (because of small sample size and limited statistical power), the observed data are more consistent with a sublinear response that approaches zero at low doses rather than a linear dose response. On the other hand, the tumor data would also probably fit a linear, low-dose model because of the small number of data points in the low-dose region.

EPA Evaluation of Bioassay Data to Estimate the CSF

For the purpose of estimating a CSF for dioxin based on animal data, EPA considered the assays conducted by Kociba et al. (1978) and NTP (1982a). In each case, EPA restricted attention to those tumor types for which incidence increased with dioxin exposure (five types in the Kociba et al. study and eight types in the NTP study). Based on an analysis by Portier et al. (1984) using a simple multistage model (order up to 3), the ED_{01} body burdens for these 13 dose-response relationships ranged from 14 to 1,190 ng/kg. The corresponding LED_{01} values ranged from 10 to 224 ng/kg.

EPA also considered two alternative ED_{01} estimates developed using the Kociba et al. (1978) female rat liver tumor data. First, EPA described an ED_{01} estimate calculated from a model developed by Portier and Kohn (1996). That model combined a pharmacokinetic model characterizing gene expression induced by dioxin with a two-stage carcinogenesis model to analyze the female rat liver tumors from the Kociba et al. study. Using that model, EPA calculated ED_{01} = 2.7 ng/kg. Second, EPA reported an estimate for ED_{01} equal to 31.9 ng/kg (LED_{01} = 22.2 ng/kg). EPA developed this estimate using its benchmark dose software and a reevaluation of the Kociba et al. (1978) pathology results by Goodman and Sauer (1992). The revised estimate also reflected other changes in the procedure for fitting a function to the data.

In 2003, when EPA's Reassessment was issued, the most recent NTP bioassay results (NTP 2004) were not yet published. Because this study represents an extensive data set developed using state-of-the-art methodology, EPA should integrate this information into its analysis.

In addition, EPA should specify criteria used to identify those data sets to be included in its analysis. The EPA Reassessment does not explain why EPA chose to rely on a single site (liver) from one sex (female) in one species (rat), as measured in a single study (Kociba et al. 1978). Consideration of other data sets that were available to EPA would have yielded a substantially wider range of potency estimates. Whereas the LED_{01} for liver tumors in female rats from the Kociba et al. study is 22.2 ng/kg, the ED_{01} values calculated using all the animal bioassay data sets considered by EPA (data sets from the NTP and Kociba et al. studies that suggested tumor incidence increases with dose) range as high as 1,190 ng/kg. In addition, use of the mechanistic model developed by Portier and Kohn (1996) to analyze the Kociba et al. female rat liver tumor data yields a substantially lower ED_{01} value (2.7 ng/kg). The range of values would be even broader if EPA had also estimated upper ED_{01} (UED_{01} [upper confidence bound on the effective dose]) values. Like the LED_{01} values, these values are indicative of the range of estimates that are consistent with the data and hence are indicative of inherent uncertainty. Calculations of a slope factor that considers the effects of dioxin on ALL tumor sites, as was done for the human epidemiological data, would probably further broaden the range of plausible ED_{01} values. Because dioxin is presumed to promote tumor growth at a wide range of sites, EPA should explain why it chose not to evaluate the dose-response relationship for "all tumors combined" in the animal studies if it considers this approach to be appropriate for use in human epidemiological studies.

The committee notes that extrapolation of results across species is highly uncertain, even when dose is scaled to account for body burden. Although data from animal and human cells and tissues suggest a qualitative similarity across species in the response to DLCs (Reassessment, Part III, p. 2-3, lines 28 and 29, and p. 3-10, lines 30 to 33), they do not support the hypothesis that the responses across species are quantitatively similar. For example, there is no explanation for the observation that the LD_{50} (50% lethal dose) for dioxin in guinea pigs and hamsters differs by more than a factor of 8,000 (Part II, p. 3-1) even though their respective receptors do not differ substantially in terms of dioxin binding and other responses (e.g., CYP1A1 induction differs by only a factor of 4). Similarly, whereas the LD_{50} for dioxin in two strains of rats differs by a factor of 300 to 1,000 (Part II, pp. 3-1 to 3-3), AHR ligand binding in these two strains has similar affinities, and CYP1A1 inducibility does not differ. These observations complicate interspecies comparisons. Recent studies comparing the response of human hepatocytes to dioxin with that of rat and mouse hepatocytes further illustrate that quantitative extrapolation of rodent data to humans is highly uncertain (Silkworth et al. 2005).

TCDD, other dioxins, and DLCs act as potent inducers of CYP, a property that can affect both the hepatic sequestration of these compounds and their half-lives. Hepatic sequestration of dioxin may influence the quantitative extrapolation of the rodent liver tumor results because the body-burden distribution pattern in highly dosed rats would differ from the corresponding distribution in humans subject to background levels of exposure. EPA should consider the possible quantitative influence of dose-dependent toxicokinetics on the interpretation of animal toxicological data.

EPA's Characterization of Uncertainty for CSF Estimates

As part of their quantification of risk, it is important for risk assessors to provide a full characterization of the uncertainty inherent in their estimates. Although risk managers may choose to focus on conservative estimates of risk, the risk assessment must be kept distinct (NRC, 1983) and should describe the lower and upper ranges of plausibility for risk estimates. A complete characterization of a risk's uncertainty facilitates (1) comparison of that risk estimate with other risk estimates that may have the same point value but a different degree of uncertainty, (2) comparison of risks with the costs and countervailing risks associated with interventions to address the primary risk, and (3) evaluation of research needs. A number of reports have addressed the need for a comprehensive treatment of uncertainty in risk assessment, including a National Research Council (NRC 1994) report.

Part III, Chapter 5, of the Reassessment describes EPA's development of a CSF for TCDD, other dioxins, and DLCs. EPA identifies 1×10^{-3} pg TEQ/kg of body weight per day $(\text{pg/kg-day})^{-1}$ "as an estimator of upper bound cancer risk for both background intakes and intakes above background" (Part III, pp. 5-28 to 5-29). While EPA qualitatively notes many of the factors contributing to this estimate's uncertainty, the Reassessment does not adequately discuss how these factors contribute quantitatively to the underlying uncertainty. By omitting the quantitative implications of these factors, the Reassessment understates the uncertainty inherent in these estimates and overstates the consistency of the data and risk estimates across all studies.

EPA should have addressed quantitatively the following sources of uncertainty:

- Basis for risk quantification: (1) bioassay data, (2) occupational cohort data.
- Epidemiology data to use: (1) risk estimate developed with data aggregated from all suitable studies, (2) risk estimate or estimates developed using each study individually.
- Factors affecting extrapolation from occupational to general population cohorts, including differences in baseline health status, age distribution, the healthy worker survivor effect, and background exposures.
- Bioassay data to use: (1) risk estimate developed with the single data set implying the greatest risk (that is, single study, tumor site, gender), (2) risk estimate developed with multiple data sets satisfying an a priori set of selection criteria.
- Dose-response model: (1) linear dose response, (2) nonlinear dose.

- Dose metric: (1) average daily intake, (2) area under the blood concentration–time curve, (3) lifetime average body burden, (4) peak body burden, (5) other.
- Dose metric—biological measure: (1) free dioxin, (2) bound dioxin.
- POD: (1) ED_{10}, (2) ED_{05}, (3) ED_{01}.
- Value from ED distribution to use: (1) ED, (2) lower confidence bound value for the ED (LED), (3) upper confidence bound for the ED (UED).

Where alternative assumptions or methodologies could not be ruled out as implausible or unreasonable, EPA could have estimated the corresponding risks and reported the impact of these alternatives on the risk assessment results. The potential impacts of four sources of uncertainty are discussed below.

- The full range of plausible parameter values for the dose-response functions used to characterize the dose-response relationship for the three occupational cohort studies selected by EPA (Ott and Zober 1996; Becher et al. 1998; Steenland et al. 2001).
- Use of other points of departure, not just the ED_{01} (or LED_{01}), to develop a CSF.
- Alternative dose-response functional forms as well as goodness of fit of all models, especially at low doses.
- Uncertainty introduced by estimation of historical occupational exposures.

If these factors are considered, the range of plausible CSF values becomes much larger, with more extreme upper and lower bound estimates, as shown below.

EPA's development of a CSF value emphasizes the analysis of three occupational cohort studies (Ott and Zober 1996; Becher et al. 1998; Steenland et al. 2001). In all cases, the studies estimated SMRs or rate ratios (RRs) as a function of cumulative dioxin lipid burden (CLB, nanogram of dioxin per kilogram of lipid weight × years) (see Reassessment, Part III, Table 5-2). Using these dose-response relationships, EPA summarizes its ED_{01} and CSF calculations (Part III, Table 5-4). The ED_{01} values are reported as lifetime average body burdens for dioxin (LABB, ng/kg). Although the relationship EPA used to convert from CLB to LABB was not transparent in the Reassessment, the committee assumes it can be described as CLB = 4 × 75 × LABB. Here, the factor of 4 accounts for conversion from nanogram per kilogram of lipid to nanogram per kilogram of body weight (see Reassessment note (a) of Table 5-

4 in Part III for this assumption), and 75 corresponds to the "average" lifetime in years.

We note that EPA identified the dose corresponding to 1% excess risk (the ED_{01}) from the relationship.

$$\frac{\text{Risk }(ED_{01}) - \text{Risk (Background)}}{\text{Risk (Infinity)} - \text{Risk (Background)}} = 0.01.$$

EPA estimated the risk function in the above equation from the occupational cohort studies by converting the hazard function to a probability of death by age 75 years. This risk estimate satisfies the requirement that risk (infinity) = 1—that is, as dose increases, the risk approaches 100%. Critical findings are reproduced in Table 5-5.

The committee also notes that EPA's analyses of the Hamburg and BASF cohorts considered all cancer deaths with no latency, but the analysis of the National Institute of Occupational Safety and Health (NIOSH) cohort considered a 15-year lag. The committee is aware of the problem with cancer mortality studies in which the subjects are without cancer at baseline (first exposure) so that the cancer mortality at the start of follow-up (in the few years after first exposure) will be artificially low. There is thus good reason to consider deaths only after some fixed time, and in dose-response calculations, one has to estimate cumulative dose appropriate to the date of occurrence of the cancer. These considerations were not part of the basis for determining the latency used in the NIOSH analysis, which was based on the assumption that effects should not be seen for many years after first exposure and the dose calculations ignored all doses for the immediately preceding 15 years. The committee is unconvinced of the validity of such assumptions in the context of dioxin as a promoter and furthermore sees no justification for considering the NIOSH results any differently than the other two cohorts.

Full Range of Plausible Parameter Values

In Part III of the Reassessment, EPA makes use of only the ED_{01} and LED_{01} for the purpose of estimating a CSF. Of course, a more complete range of plausible CSF values can be developed by considering parameter estimates corresponding to dose-response relationships that are less than the central estimate relationship (used to identify the ED_{01}). EPA's recently released cancer guidelines (EPA 2005a) recommends use of both lower- and upper-bound values. In section 3.2.4 of the document, which is entitled Point of Departure (POD), EPA states, "risk assessors should calculate, to the extent practicable, and present the central estimate and the correspond-

TABLE 5-5 EPA Inputs to CSF Estimates Using Epidemiological Data[a]

Study	Function RR(x)[b]	Central Estimate for b[b]	P Value for b[b]	ED_{01}[c]	LED_{01}[c]
Becher et al. 1998	Power: $(1 + kx)^b$	$b = 0.326$ for $k = 1.7 \times 10^{-4}$	0.026	6	NA[d]
	Additive: $1 + bx$	$b = 1.6 \times 10^{-5}$	0.031	18.2	NA[d]
	Multiplicative: e^{bx}	$b = 8.69 \times 10^{-6}$	0.043	32.2	NA[d]
Steenland et al. 2001	Power: $(x/\text{background})^b$	$b = 0.097$	0.003	1.38[e]	0.71[e]
	Piecewise linear: e^{bx}	$b = 1.5 \times 10^{-5}$	NA[f]	18.6	11.5
Ott and Zober 1996	e^{bx}	$b = 5.03 \times 10^{-6}$	0.05	50.9	25

[a]ED_{01} values here represent estimates only for males (as is the case in EPA's Part III, Table 5-4). Female ED_{01} values are modestly larger because the background cancer rate for females is less than it is for males. For example, the Ott and Zober (1996) ED_{01} value for females is 62 ng/kg and the corresponding LED_{01} is 30.5 ng/kg.

[b]See Part III, Table 5-2. Note that x is exposure expressed in terms of cumulative lipid burden (CLB), ng of dioxin/kg of fat × years.

[c]See Part III, Table 5-4. The ED_{01} values are expressed in terms of lifetime average body burden (LABB), ng of dioxin/kg of body weight.

[d]Not available. EPA did not estimate LED_{01} values (or the corresponding upper bound for ED_{01}), although these values can be calculated, as described below.

[e]EPA reported these values in Part III, Table 5-3. Note that EPA omitted further consideration of the power function for this dataset, stating that "this formula predicts unreasonably high attributable risks at background dioxin levels in the community due to the steep slope of the power curve formula at very low levels" (Part III, p. 5-37).

[f]Not available. Steenland et al. (2001) did not report the *P* value for this parameter, although EPA reported a value for LED_{01}.

ing upper and lower statistical bounds (such as confidence limits) to inform decision makers" (EPA 2005a, p. 3-17).

To illustrate the quantitative impact on the range of uncertainty from this one assumption, the committee considered the upper ED_{01} (UED_{01}) values that correspond to the lower 95% confidence interval on the dose-response relationship. EPA provides the UED_{01} values for the Steenland et al. (2001) and Ott and Zober (1996) studies (see Table 5-3 in Part III of the Reassessment). As explained below, the committee has calculated the UED_{01} values for the Becher et al. (1998) study. Together with the ED_{01} and LED_{01} values, the UED_{01} values help to describe the range of plausible ED_{01} values and hence the uncertainty that attends the CSF estimates due to finite sampling.

To estimate the range of plausible ED_{01} values, the committee assumes that the set of plausible values for the dose-response relationship parameter

(b—see column 3 of Table 5-5 in this report) is normally distributed with a mean equal to the parameter's central estimate (b—see column 3 of Table 5-5) and a standard deviation equal to the estimate's standard error (b_{SE}).[4] When possible, the committee estimated the value of b_{SE} using the P value for the dose-response relationship parameter, as reported in the Reassessment, Part III, Tables 5-2 and 5-3 (shown in column 4 of Table 5-5 of this report). In particular, b_m, b_{SE}, and P satisfy the relationship

$$b_m - N^{-1}\left(1 - \frac{P}{2}\right) \times b_{SE} = 0,$$

where N^{-1} is the inverse cumulative normal function. For example, if b_m = 0.1 and $P = 0.05$, then $b_{SE} = 0.051$. Designating $b_{UED} = b_m - 1.96b_{SE}$, the value of b yielding the UED_{01}, and $b_{LED} = bm + 1.96b_{SE}$, the value of b yielding the LED_{01}, the UED_{01} satisfies the relationship $RR(b_{UED}, UED_{01}) = RR(b_{LED}, LED_{01})$, where the function RR is the dose-response relationship (rate ratio) taking two arguments (the parameter b and a dose).

When the P value is not available but both LED_{01} and ED_{01} are specified, the committee assumed that $b_{LED} - b_m = b_m - b_{UED}$. If necessary, the value of b_{LED} was estimated from the relationship $RR(b_{LED}, LED_{01}) = RR(b_m, ED_{01})$ and UED_{01} from the relationship $RR(b_{UED}, UED_{01}) = RR(b_{LED}, LED_{01})$. In the case of the Becher et al. (1998) study, inserting b_m into any of the RR formulas along with the ED_{01} value for that dose-response function yields an RR of approximately 1.09. It was assumed that the UED_{01} is the dose that yields an RR of 1.09 when inserted into the dose-response function along with b_{UED}. For example, the Becher et al. power function yields an RR of 1.09 if a LABB of 6 ng/kg (CLB = 1,800 ng/kg-year) is used along with the exponent parameter $b_m = 0.326$. In particular, $(1 + 0.00017 \times 1{,}800)^{0.326} = 1.09$. The value of b_{UED} is 0.039 and $(1 + 0.00017 \times 45{,}300)^{0.039} = 1.09$. That is, CLB = 45,300 ng/kg-year produces the same RR when used with $b = b_{UED}$. Dividing CLB by $4 \times 75 = 300$ yields a LABB of 151 ng/kg. The committee assumed that because the LABB of 151 ng/kg also yields an RR of 1.09, this dose is the UED_{01}.

Table 5-6 summarizes the ED_{01}, LED_{01}, and UED_{01} values for the dose-response relationships listed in Table 5-2 of Part III of the Reassessment. The results in Table 5-6 indicate that the set of plausible ED_{01} values spans at least one or two orders of magnitude for the Becher et al. (1998) study and the Ott and Zober (1996) study.

---

[4]When estimated using a large number of observations, statistical parameters typically have normal error distributions. Of course, it is possible that the error distributions for the b_m parameters are not normal and hence not symmetric.

TABLE 5-6 ED_{01}, LED_{01}, and UED_{01} Values

		LABB (ng/kg)		
Study	Function RR =	ED_{01}	LED_{01}	UED_{01}
Becher et al. 1998	Power: $(1 + kx)$	6	3	**150**
	Additive: $1 + bx$	18.2	9.5	**200**
	Multiplicative: e^{bx}	32.2	16	**1,000**
Steenland et al. 2001	Power: (x/background)	1.38	0.71	8.95
	Piecewise linear: e^{bx}	18.6	11.5	49
Ott and Zober 1996	e^{bx}	50.9	25	Infinite

NOTES: Bolded values represent the committee's estimates of UED_{01} for the Becher et al. (1998) study. These values were estimated by assuming a normal error distribution for b (see column 3 in Table 5-5 of this report). All other values were as reported by EPA in Table 5-3 of Part III of the Reassessment.

Consideration of Alternative Points of Departure

EPA explains that while a 10% level is generally used as a POD (that is, an ED_{10} is generally used to estimate the CSF), "where more sensitive data are available, a lower point for linear extrapolation can be used to improve the assessment (e.g., 1% response for dioxin, ED_{01})" (Part III, p. 5-15). EPA's cancer guidelines (EPA 2005a, see also Appendix B) state, "Conventional cancer bioassays, with approximately 50 animals per group, generally can support modeling down to an increased incidence of 1-10%; epidemiologic studies, with larger sample sizes, below 1%" (p. 3-17).

However, these generalities do not imply that extrapolation down to low levels is justified in all circumstances. EPA's carcinogen risk assessment guidelines document explains, "Various models commonly used for carcinogens yield similar estimates of the POD at response levels as low as 1% . . . . Consequently, response levels at or below 10% can often be used as the POD" (EPA 2005a, p. 3-17). The key point here is that a lower response level is justified only if the estimated dose corresponding to this response is insensitive to the functional form (provided the other functional forms fit the data to a comparable degree). The dose-response functions for the epidemiological data identified by EPA suggest this criterion is not satisfied. For example, as detailed in Table 5-3 of Part III of the Reassessment, the ED_{01} for males in the Steenland et al. (2001) study is 1.38 ng/kg of body burden if the power function is used, more than an order of magnitude less than the ED_{01} of 18.6 ng/kg calculated using the piecewise linear function. In the Becher et al. (1998) study, the ED_{01} spans a factor of five, depending on which dose-response function is used.

Although EPA states that a 1% response above background (corresponding to RR ≈ 1.09) is within the range of observed response for the three occupational cohort studies considered, it is clearly at the low end of the observed range. For example, among the five exposure groups defined in the Becher et al. (1998) study (excluding the comparison group, for which SMR is fixed at 100), the lowest RR is 1.12. Of the six exposure groups in the Steenland et al. (2001) study (excluding the comparison group), one has an RR value below 1.09 (RR = 1.02 for the second lowest exposure group). RR values for the other five groups were 1.26 or greater. For the Ott and Zober (1996) study, RR values for the three comparison groups were 1.2, 1.4, and 2.0.

The use of alternative points of departure for the power dose-response relationships would greatly increase the range of plausible CSF values. Table 8-2 (Reassessment, Part II) demonstrates this point. For the Steenland et al. power function, the 95% confidence interval for ED_{01} spans approximately one order of magnitude. As a result, the CSF calculated with this set of ED_{01} estimates likewise spans approximately an order of magnitude. In contrast, the 95% confidence interval for ED_{05} spans approximately three orders of magnitude. Similarly, the ED_{01} confidence interval derived from the Becher et al. power function spans a factor of approximately 50. The corresponding range for ED_{05} spans nearly four orders of magnitude.

Thus, it is evident that the choice of POD can have a substantial impact on the uncertainty of the final risk estimate, especially if both upper and lower confidence limits are provided. The importance of this assumption is not readily evident in the Reassessment. The transparency of the uncertainty of CSF calculations, and thus risk estimates, would be substantially improved if the document presented CSF ranges and risk estimates calculated from both ED_{01} and ED_{05} values to illustrate the importance of this assumption.

Consideration of Alternative Dose-Response Functional Forms

Because there are so many functional forms from which to choose for the purpose of modeling dose response, EPA should establish criteria for selecting acceptable solutions. For example, there are formal goodness-of-fit tests that can help to identify the best candidates. Note that a higher statistical significance for a positive dose response does not necessarily imply that, using standard statistical criteria, the model adequately fits the data. Evaluating the goodness of fit for the occupational cohort analyses was complicated by EPA's lack of ready access to the original data. Despite these complications, it is important that EPA provide a cogent set of criteria for determining which functional forms were used. This section identifies four instances in which EPA eliminated from consideration alternative dose-response functional forms without providing adequate justification. EPA should describe the range of ED_{01} and ED_{05} values implied by dose-response functions that are statistically consistent with the occupational cohort data and the inclusion criteria established for this assessment.

First, EPA eliminated from consideration the power function dose-response relationship calculated from the Steenland et al. (2001) study, explaining only that this relationship "predicts an unrealistic risk for the background exposure" (Reassessment, Part II, p. 8-67) and that it "leads to unreasonably high risks at low exposure levels, based on calculations of the attributable risk that this model would predict from background dioxin levels in the general population" (Reassessment, Part III, p. 5-13). EPA provided no criteria by which it judged the reasonableness of the Steenland et al. power function, nor does EPA provide any further explanation on this point. The Reassessment should provide further scientific rationale for excluding the Steenland et al. power function, or it should be considered as valid as any of the other dose-response relationships.

Second, EPA considered only dose-response relationships based on the assumption of no background incremental risk (that is, SMR = 100 at background exposure levels). This assumption is inconsistent with the findings of two analyses identified by EPA (Starr 2001, 2003; Crump et al. 2003) (Part III, p. 5-14) that rejected the assumption on statistical grounds that SMR = 100 at baseline exposure levels. If relaxing this assumption yields an estimate of SMR > 100 at background exposures, the resulting dose-response relationship would tend to be shallower, yielding smaller CSF values. For example, EPA (Part III, p. 5-15) noted that in a pooled analysis of Ott and Zober (1996), Flesch-Janys et al. (1998), and Steenland et al. (2001), fixing SMR = 100 at background exposure levels yielded ED_{01} = 51 ng/kg-day, whereas dropping this assumption resulted in ED_{01} = 91 ng/kg. EPA should provide an explanation for assuming SMR = 100 at background exposure levels. Short of doing so, EPA should consider the impact of relaxing this assumption on the estimated value of the ED_{01}.

Third, for the piecewise linear dose-response function developed for the Steenland et al. data set, EPA considered only one cut point (40,000 ng/kg × years) (the cut point, or changing point, is the dose at which the slope of the piecewise linear dose-response relationship changes). Although this is the best-fit cut-point estimate and the only relationship of this form reported by the authors, other cut-point values are plausible. (Other cut points would yield dose-response relationships that could not be statistically rejected.)

Finally, EPA considered only a subset of the plausible dose-response relationships that could be fit to the data in the Becher et al. (1998) study. Becher et al. considered a family of dose-response relationships of the form $RR = (kx + 1)^{\beta}$, where the value of k is chosen arbitrarily. The best-fit value of β depends on the value of k selected. The relative plausibility of different values of k can be determined by comparing likelihood function values. Finally, holding $\beta = 1$ yields a linear function where the value of k is uncertain. Becher et al. reported that $k = 0.00017$ maximizes the likelihood function and that, for this value of k, the corresponding value of $\beta = 0.326$. Holding $\beta = 1$ yields $k = 0.000016$. However, Figure 1 of Becher et al. indicates that the likelihood function is relatively insensitive to the value of k selected. Hence, other dose-response relationships are plausible. Estimating the ED_{01} values corresponding to these alternative dose-response relationships would require further primary analysis of the data.

Uncertainty Associated with Estimation of Historical Exposures

The assumed half-life for dioxin in humans plays a major role in the back-extrapolation of dioxin lipid concentrations to the estimation of peak body burdens in occupational cohorts. The Reassessment states, "Using published first-order back-calculation procedures, the relatively small difference (<10-100-fold) in body burden between exposed and controls in the dioxin epidemiology studies makes exposure characterization in the studies a particularly serious issue" (Part III, p. 5-7). The high exposures in the occupational cohorts suggest a high likelihood of enzyme induction during the period of occupational exposure that may have led to a reduction of the half-life to less than the assumed value of 7.1 years. Aylward et al. (2005) discussed the issue of half-lives and the impact of this parameter on risk estimates.

EPA's Reassessment compared the impact of using either a half-life of 4 years or the default of 7.1 years on the back-extrapolation estimate. EPA reported that using a 4-year half-life increases the peak body burden and the area under the curve (AUC) by 4.6-fold and 3.8-fold, respectively. This difference would have increased the estimated ED_{01} values by the same amount and hence decreased the CSF estimates, resulting in a lower risk estimate. Given the potential importance of this issue, the committee finds the following statement by EPA surprising: "This bounding exercise suggests that impacts on back-calculated peak and AUC values may become significant if the models predict prolonged periods with half-lives of less than 4 years" (Part III, p. 5-8).

Because the impact of the half-life used for back-extrapolation depends on the back-extrapolation duration required in any particular study, EPA should have estimated the impact of using the 4-year alternative value for each of the main epidemiological studies separately. EPA should also consider the issues raised by Aylward et al. (2005). Overall, the Reassessment does not provide sufficient quantification of the impacts of these choices, and the committee believes these decisions influence the estimated dose-response relationships.

Overall Uncertainty

Table 5-4 (Reassessment, Part III) summarizes EPA's all-site cancer ED_{01} values. For the three occupational cohort studies, these values span less than an order of magnitude (6 to 50.9 ng/kg). The corresponding CSF values range from 5.7×10^{-4} to 5.1×10^{-3} $(pg/kg\text{-}day)^{-1}$. On the basis of this range, EPA concludes that "A slope factor estimate of approximately 1×10^{-3} per pg of dioxin per kg of body weight per day represents EPA's most current upper bound slope factor for estimating human cancer risk based on human data" (Part III, p. 5-28). The animal-based ED_{01} values listed in EPA's Table 5-4 range from 22 to 30.9 ng/kg, leading to the conclusion that "A slope factor of 1.4×10^{-3} per pg dioxin/kg body weight/day represents EPA's most current upper bound slope factor for estimating human cancer risk based on animal data" (Part III, p. 5-28).

Whereas a CSF of approximately 1×10^{-3} per pg/kg-day (equivalently, an ED_{01} of approximately 30 ng/kg LABB) lies within the range of plausible values, this discussion has focused on the relative magnitude of the range of plausible values. Consideration of the set of all plausible parameter values (that is parameter values within the 95% confidence interval) for the dose-response functions considered by EPA considerably widened the range of values estimated from the Becher et al. and Ott and Zober studies (see Table 5-6). The CSF values (risk per pg/kg-day) can be calculated by first converting the ED_{01} expressed as ng/kg LABB to an ED_{01} expressed as a daily intake (ng/kg-day) using EPA equation 5-1 (Part III, p. 5-18) and then dividing this intake into a risk of 0.01. For the Becher et al. (1998) study, the resulting CSF values range from 3.0×10^{-5} to 1.0×10^{-3} per pg/kg-day, more than two orders of magnitude. The 95% confidence interval for the CSF calculated from the Ott and Zober study (1996) has a lower bound of zero and an upper bound of 1.2×10^{-3}. Only the Steenland et al. (2001) study retains an ED_{01} range (CSF range) with a span confined to less than two orders of magnitude (CSF 6.1×10^{-4} to 3.0×10^{-2}). Figure 5-2 compares the range of plausible CSF values identified by EPA with the range of plausible values consistent with the dose-response parameter 95% confidence intervals.

Consideration of alternative points of departure can greatly inflate the confidence interval for the power function dose-response relationships. Using an ED_{05} (rather than an ED_{01}) broadens the confidence interval for the function to more than three orders of magnitude. Consideration of alternative dose-response relationship forms could further broaden the range of plausible CSF values, although primary analysis of the data would be required to quantify the impact. The Reassessment could develop a distribution for the CSF by assigning some probability to different options for each of the assumptions discussed here (McKone and Bogen 1992; Evans et al.

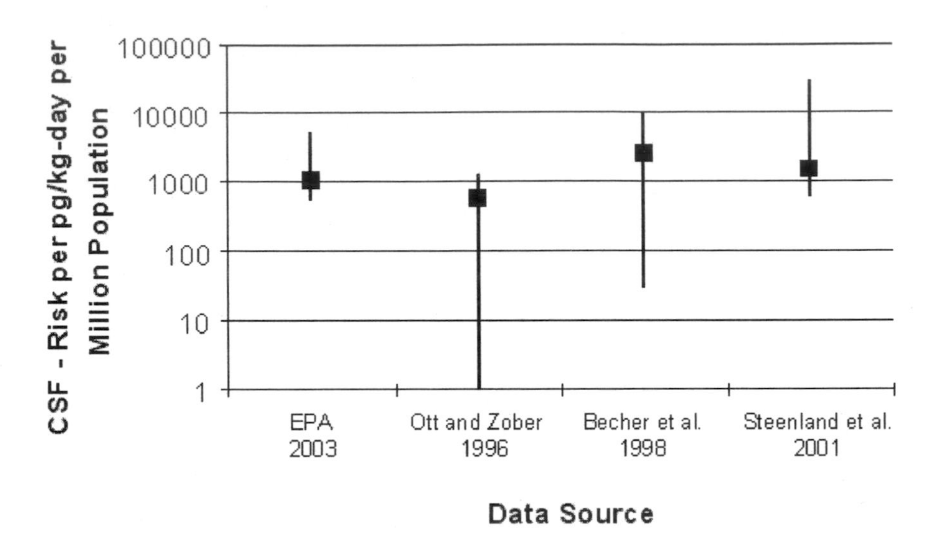

FIGURE 5-2 Range of plausible CSF values: consideration of parameter confidence intervals only. Solid blocks are central estimate values. For EPA, this value (1,000) corresponds to EPA's stated central estimate of dioxin's CSF (1×10^{-3} per pg/kg-day). For the Becher et al. (1998) study, the central estimate in the figure corresponds to the average of the three ED_{01} values that EPA reports in Part III, Table 5-4. For the Ott and Zober (1996) and Steenland et al. (2001) studies, the central estimate corresponds to the individual ED_{01} values listed in EPA, Part III, Table 5-4.

1994a,b). In any case, a more thorough consideration of plausible alternative values for key assumptions is needed to portray accurately to risk managers the magnitude of uncertainty that underlies the quantitative risk estimates derived from epidemiological studies.

CONCLUSIONS AND RECOMMENDATIONS

Qualitative Weight-of-Evidence Carcinogen Classification

• The committee concluded that the classification of dioxin as "carcinogenic to humans" versus "likely to be carcinogenic to humans" depends greatly on the definition and interpretation of the specific criteria used for classification, with the explicit recognition that the true weight of evidence lies on a continuum with no bright line that easily distinguishes between these two categories. The committee agreed that, although the weight of epidemiological evidence that dioxin is a human carcinogen is not strong, the human data available from occupational cohorts are consistent with a modest positive association between relatively high body burdens of dioxin and increased mortality from all cancers. Positive animal studies and mechanistic data provide additional support for classification of dioxin as a human carcinogen. However, the committee was split on whether the weight of evidence met all the necessary criteria described in the cancer guidelines (EPA 2005a, see also Appendix B) for classification of dioxin as "carcinogenic to humans." EPA should summarize its rationale for concluding that dioxin satisfies the criteria set out in the most recent cancer guidelines (EPA 2005a, see also Appendix B) for designation as either "carcinogenic to humans" or "likely to be carcinogenic to humans."

- The committee agreed that other DLCs are most appropriately classified as "likely to be carcinogenic to humans."
- Should EPA continue to classify dioxin as "carcinogenic to humans," more justification will be required to rationalize why a mixture containing dioxin would not also meet the classification of "carcinogenic to humans."
- If EPA continues to designate dioxin as "carcinogenic to humans," it should explain whether this conclusion reflects a finding that there is a strong association between dioxin exposure and human cancer or between dioxin exposure and a key precursor event of dioxin's mode of action (presumably AHR binding). If EPA's finding reflects the latter association, EPA should explain why that end point (e.g., AHR binding) represents a "key precursor event."
- The committee considers the distinction between these two categories to be based more on semantics than on science and recommends that EPA spend its energies and resources more carefully delineating the assumptions used in quantitative risk estimates for TCDD, other dioxins, and DLCs derived from human and animal studies.

Quantitative Risk Estimation of Cancer Potency

- The committee concludes that there is an adequate scientific basis to support the hypothesis that the shape of the relationship between dioxin dose and cancer risk is sublinear at low doses, perhaps reflecting responses indistinguishable from background risk at doses below which dose-response data are available, including evidence that (1) TCDD, other dioxins, and DLCs are not genotoxic; (2) dioxin acts through receptor mediation, and receptor-mediated carcinogens tend to exhibit sublinear dose-response relationships; (3) dioxin-induced liver tumors are secondary to hepatotoxicity and enhanced rates of cell proliferation; (4) bioassay results suggest sublinearity (Hill coefficient central estimates substantially greater than 1); and (5) epidemiological results do not help to distinguish between zero and nonzero responses at the low-dose end of the dose-response curve. Accordingly, a risk assessment can be conducted without resorting to default assumptions.

• To the extent that EPA favors using default assumptions for regulating dioxin as though it were a linear carcinogen, such a conclusion should be supported with scientific evidence. (For example, EPA could explore whether background exposures raise the population to the linear portion of the dose-response relationship.) Alternatively, the decision to use the linear dose-response relationship could be made as a part of risk management, although the risk assessment should provide the scientific strengths and weaknesses for both linear and nonlinear approaches. EPA should adhere to the division between risk assessment, which is a scientific activity, and risk management, which takes into account other considerations, as described by the National Academy of Sciences more than two decades ago (NRC 1983).

• EPA has not adequately justified use of the 1% excess risk level as the POD for the analysis of either the epidemiological or animal bioassay data. Although demonstrating that the POD is within the range of the data is necessary, it is not sufficient to justify its use. Other conditions, such as demonstrating that the POD is relatively insensitive to functional form (as noted in EPA's cancer guidelines), must also be satisfied. EPA should acknowledge the larger extrapolation from justifiable POD values down to environmentally relevant doses that would be necessitated by use of a higher-response-level POD.

• Regarding EPA's review of the animal bioassay data, the committee recommends that EPA establish clear criteria for including different data sets. The reliance on one site from one gender of one species, as reported by a single study, does not adequately represent the full range of data available. The committee recommends that EPA consider the full range of data, including the new NTP animal bioassay study on TCDD for quantitative dose-response assessment.

Characterization of Uncertainty Surrounding Cancer Risk Estimates

• EPA should characterize more completely the uncertainty associated with risk estimates inferred from the epidemiological data by (1) taking into account the full range of ED_{xx} values statistically consistent with the data (not just the central and lower estimates), (2) considering alternative PODs, (3) considering alternative dose-response functional forms consistent with the data, and (4) considering uncertainty associated with the half-life estimates of dioxin in humans for the purpose of back-extrapolating exposures in occupational cohort studies.

• The committee recognizes that explicit characterization of uncertainty could result in an especially wide range of risk estimates. Narrowing consideration to a subset of those estimates could be made as part of the risk management task. For example, a "health protective" (conservative) estimate could be used to support an imperative to protect public health. Alternatively, if the goal is to compare that risk with other risk management priorities, countervailing risks, or the economic costs of risk mitigation, a central or arithmetic mean value could be used. Finally, to address uncertainty associated with specification of the dose-response relationship functional form below the POD (that is, linear vs. nonlinear), EPA could choose to use a margin of exposure approach in place of estimating population risk. These options are the purview of risk management rather than risk assessment.

- On the whole, it was the committee's impression that EPA's narrative in discussing epidemiological studies in Part III of the Reassessment tended to focus on positive findings without fully considering the strengths and limitations of both positive and negative findings. Part III of the Reassessment would be strengthened if EPA clearly identified specific inclusion criteria for those studies for which quantitative risk estimates were determined.

6

Noncancer End Points

This chapter reviews the U.S. Environmental Protection Agency (EPA) assessment of the noncancer end points, including immune function, reproduction and development, diabetes, thyroid function, lipid levels, and other effects related to exposure to 2,3,7,8-tetrachlorodibenzo-*p*-dioxin (TCDD, also referred to as dioxin), other dioxins, and dioxin-like compounds (DLCs) in animals and humans. The purpose of this chapter is to critically assess, to the extent possible, whether EPA has met the criteria set forth in the "Statement of Task" with respect to the noncancer effects of TCDD, other dioxins, and DLCs. Toward this end, this chapter focuses on the uncertainties and assumptions made by EPA in determining whether TCDD, other dioxins, and DLCs have noncancer effects in humans; determining the methods and models used for assessing these effects; determining the breadth and robustness of the studies used and the balance with which the studies are presented in the Reassessment,[1] and finally, determining whether the conclusions reached by EPA are consistent with the current scientific peer-reviewed literature.

IMMUNE FUNCTION

EPA uses a sizeable immunotoxicology database derived largely from laboratory animal studies and a smaller number of epidemiological studies in

[1] *The Exposure and Human Health Reassessment of 2,3,7,8-Tetrachlorodibenzo-p-dioxin (TCDD) and Related Compounds* (EPA 2003a, Part I; 2003b, Part II; 2003c, Part III) is collectively referred to as the Reassessment.

its assessment of immunotoxicity produced by TCDD, other dioxins, and DLCs.

The assessment of changes in immune competence, regardless of the cause, is complex, as the immune system is composed of a large and diverse group of cellular and soluble components. In addition, compensatory and overlapping mechanisms in host immunity can make it difficult to identify subtle or modest changes within the immune system.

Because of the many different cell types and soluble factors, which alone or cooperatively mediate a wide variety of evocable immunological responses, there is no one test or assay that can measure all the different elements. Therefore, the approaches used to identify changes in immune status are multifaceted, including pathological examination of lymphoid organs, enumeration of leukocyte subpopulations, quantification of soluble mediators, and measurement of immune function responses, such as susceptibility to infection or reduced immunological responses to vaccines. Standard assays are available for all the above determinations; however, because of the sheer enormity of the task to measure them all, immune competence is typically assessed by using a small number of immunological end points, often quantifying various aspects of innate, humoral, and cell-mediated immunity. Therefore, the EPA review draws on a large but diverse database of studies that are often difficult to compare because different assays, model systems, responses, and animal species were used.

EPA draws several important conclusions about the immunotoxicity of TCDD, other dioxins, and DLCs that are summarized in the Reassessment, Part III, Integrated Summary and Risk Characterization. The first is that "there appears to be too little information to suggest definitively that 2,3,7,8-TCDD at levels observed (in the reported studies) causes long-term adverse effects on the immune system in adult humans" (p. 2-34; lines 19 to 21). The second is that "cumulative evidence from a number of studies indicates that the immune system of various animal species is a target for toxicity of TCDD and structurally-related compounds, including PCDDs, PCDFs [polychlorinated dibenzofurans] and PCBs [polychlorinated biphenyls]" (p. 2-34; lines 24 to 26). Third, animal studies show that TCDD suppresses both "cell-mediated and humoral immune responses, suggesting that there are multiple cellular targets within the immune system that are altered by TCDD" (p. 2-34; lines 26 to 28). EPA goes on to state that "it can be inferred from the available data that dioxin-like congeners are immunosuppressive" in animals. Fourth, the weight of evidence from animal studies in vivo and in vitro "supports a role for Ah-mediated immune suppression" by DLCs (p. 2-35, lines 27 to 28); "other in vivo and in vitro data, however, suggest that non-AHR (aromatic hydrocarbon receptor)-mediated mechanisms may also play some role in immunotoxicity" (p. 2-35, lines 28 to 30). Finally, EPA concludes that "there are insufficient clinical data from these studies to fully assess human sensitivity to TCDD exposure. Nevertheless, because of extensive animal work, the database is sufficient to indicate that immune effects could occur in the human population from exposure to TCDD and related compounds at some dose level. At present, it is EPA's scientific judgment that TCDD and related compounds should be regarded as nonspecific immunosuppressants and immunotoxicants until better data to inform judgment are available" (p. 2-37, lines 4 to 10). The strengths and weaknesses of those conclusions are discussed below.

Uncertainties and Assumptions in Determining Whether TCDD Is Immunotoxic in Humans

Is the Assumption Correct That the Immune System in Humans and in Animal Models, Primarily Mice, Are Similar?

Historically, the mouse has been the animal model of choice for immunologists; it has also been widely embraced as the model of choice for immunotoxicological studies. Hence, most immunotoxicological studies used by EPA in preparing its report are based on studies in mouse models. The human and mouse immune systems are similar in composition and function; therefore, from the standpoint of comparisons based on the composition of this target organ, it is reasonable to assume that studies in mice provide important qualitative insights into the mechanism of action of TCDD, other dioxins, and DLCs on the human immune system. Providing, insights into the mechanism of action is one of the primary strengths of the mouse model for which there are genetically defined AHR high- and low-responding mouse strains, congenic mice at the Ahr locus, and AHR null ($Ahr^{-/-}$) mice. More reagents, assays, and biological probes are available for the mouse immune system than for any other species except the human immune system. However, as is the case for other toxicological end points, information-derived from animal studies is qualitative in that the pharmacodynamics, pharmacokinetics, half-lives of the compounds, affinity of AHR, linkage of the receptor to signal transduction pathways, and numerous other factors can be significantly different, at least quantitatively, between humans and other animal species. Therefore, direct quantitative extrapolations from animal models to humans can result in a significant underestimation or overestimation of risk. When strong scientific evidence exists concerning specific species differences, these factors should be incorporated into the risk characterization.

Is the Toxic Equivalency Factor/Toxic Equivalent Quotient Approach for Estimating the Immunotoxicity of Mixtures Scientifically Justified?

The toxic equivalency factor/toxic equivalent quotient (TEF/TEQ) approach is based on a well-defined structure-activity relationship for persistent dioxins, other than TCDD, and DLCs for which there is a positive correlation between AHR affinity and toxic potency. There is good general agreement in comparisons of results across studies, primarily in mice, where the acute immunotoxic effects for individual dioxin-like congeners were examined, suggesting that the immunotoxic potency for various congeners and AHR activation exhibit the same qualitative rank order. Few immunotoxicological studies investigated structure-activity relationships for immunotoxic potency. Of those studies, with only a few exceptions, the rank order immunotoxic potency correlated positively with AHR activation. Several exceptions to this relationship are found in the current literature, and some are discussed in the Reassessment, Chapter 4, Part II. For example, 2,7-dichlorodibenzo-*p*-dioxin, a congener that would be predicted to exhibit low binding affinity for AHR, was found to be equipotent in suppressing the anti-sheep red-blood-cell (anti-SRBC IgM) antibody forming response to TCDD. For this example, the TEF/TEQ approach would not provide a reasonable estimate of immunotoxicity. Because in vivo immunotoxicity data for 2,7-dibenzo-*p*-dioxin are available only in the mouse, it is unclear whether the unexpected potency of this congener occurs in other animal species and humans. A second example pertains to certain halogenated aromatic hydrocarbons (HAHs), including several diortho PCB congeners, which exhibit antagonist activity when administered in a mixture with AHR agonists. The latter point may not be trivial, as several of the diortho PCB congeners (e.g., PCB153) are abundant environmental cocontaminants with dioxins, other than TCDD, and DLCs. The aforementioned examples of exceptions for which the TEF/TEQ approach might not predict toxic potency accurately are presented and discussed in a balanced manner (Part II, pp. 4-6 to 4-7, lines 19 and 20).

In spite of the aforementioned caveats, based on what is known about the cell biology of AHR and the mechanisms of immunotoxicity for TCDD and related compouds, the TEF/TEQ approach for assessing the immunotoxic potency of mixtures of persistent dioxins, other than TCDD, and DLCs is scientifically justified. Having said that, the effective application of the TEF/TEQ approach for assessing immunotoxic potential ultimately would critically depend on the TEF values assigned to individual congeners for a given immunological response. What is unclear is which immune responses should be used in risk management, as all are not equally sensitive to modulation by HAHs.

What Is the Profile of Immune Toxicity in Humans Exposed to DLCs?

Well-documented human exposure to TCDD, other dioxins, and DLCs has occurred in the occupational setting and within the general population from industrial accidents and through consumption of contaminated food. Even though immune function and status have been examined in exposed individuals, only a small number of studies have been conducted with appropriate controls and accurate measures of exposure. Likewise, a small number of epidemiological investigations evaluating immunologically related outcomes from chronic exposures have been reported. The results from these human studies for the most part have yielded inconclusive results. In several studies, modest changes in immune status were observed; yet in other studies, the findings were not reproduced or were even contradicted. In animal studies, it is clear that TCDD, other dioxins, and DLCs markedly suppress both humoral and cell-mediated immune responses. This profile of activity has not been unequivocally demonstrated in humans. The most obvious reason for the inconclusive findings in human studies is that, in many cases, a very small number of subjects were evaluated, their primary exposures were often long before measurement of immune end points and concentrations of TCDD and related compounds, and the nature of those exposures were often unclear. Furthermore, obtaining an accurate estimate of the level of exposure through back-extrapolation from current body levels may be more complex than simply using a single half-life throughout.

Another contributing factor to the inconsistent results is that immune responses to defined stimuli are highly variable among humans. This variability can be attributed to genetic variability, age, environmental history, and other still undefined causes. This variability in individuals substantially limits the ability to identify subtle and even moderate alterations of immune function after exposure to agents, especially in human populations. Therefore, in the absence of more comprehensive immunotoxicological human data, it is reasonable to assume that TCDD, other dioxins, and DLCs will exert a profile of immunotoxicity comparable to that observed in animal models, such as mice. For risk analysis, the more critical issue is whether the immunotoxic potency observed in certain animal models is significantly greater (by orders of magnitude) than in humans.

What Is the Sensitivity of the Human Immune System to DLCs?

For many of the same reasons as discussed in the previous section, the sensitivity of humans to immune suppression by TCDD, other dioxins, and DLCs is also currently unclear. There are four separate reports of a longitudinal study in a cohort of Dutch children suggesting that the developing human immune system may be susceptible to immunotoxic alterations from exposure to Western European environmental levels of TCDD, other dioxins, and DLCs (Weisglas-Kuperus et al. 1995, 2000, 2004; ten Tusscher et al. 2003). Exposures before 1990 in the Netherlands resulted in breast milk concentrations of TEQ (at the start of the study) at 30 to 60 parts per trillion (ppt) (lipid), whereas concentrations in the United States at approximately the same time period were in the 15- to 24-ppt range. Three industrial areas were compared with a rural area with about 20% less PCBs in maternal plasma (Koopman-Esseboom et al. 1995). The Dutch study measured the major PCB congeners in plasma in the mother and the newborn and all TCDD and related compounds in maternal breast milk 10 days and 3 months postpartum (Feeley 1995). About half the study subjects were fed a formula that had TCDD and related compounds at less than 2 ppt, and the other half of the test subjects were breast-fed and could be divided into groups breast-fed less than 4 months and those breast-fed more than 4 months. Therefore, individual calculations of total exposure could be correlated with plasma PCB concentrations in children at 3 months, 18 months, 42 months, and 9 years. Three more recent studies (Weisglas-Kuperus et al. 2000, 2004; ten Tusscher et al. 2003) are important because some of the same findings observed in these later studies were also reported in the 1995 studies, indicating persistence of alterations. Weisglas-Kuperus et al. (2000) reported on 207 healthy mother-infant pairs with increased prenatal exposure to TCDD, other dioxins, and PCBs; the results showed an association between exposure and immunological changes, which included an increase in number of lymphocytes, γ-δ T cells, $CD3^{+}HLA\text{-}DR^{+}$ (activated) T cells, $CD8^{+}$ cells, $CD4^{+}CD45RO^{+}$ (memory T cells), and lower antibody levels after mumps and measles vaccination at preschool age. In addition, an association was found between prenatal exposure and decreased shortness of breath with wheeze, and current PCB burden was associated with a higher prevalence of recurrent middle-ear infections and chicken pox and a lower prevalence of allergic reactions. Although an association between TCDD and PCB exposure and changes in immune status was observed, all infants were found to be in the normal range. In a second study, ten Tusscher et al. (2003) reported modest but persistent changes in immune status in children with perinatal exposure to dioxin as evidenced by a decrease in allergy, persistently decreased thrombocytes, increased thrombopoietin, increased $CD4^{+}$ T cells, and increased $CD45RA^{+}$ cell counts in a longitudinal subcohort of 27 healthy 8-year-old children with documented perinatal exposure to TCDD. The original cohort at 42 months demonstrated an association between reduced vaccine titers, increased incidence of chicken pox, and increased incidence of otitis media with higher TEQ. However, by 8 years of age, the more frequent recurrent ear infections were still apparent (overall), although the chicken pox frequency showed an inverse correlation with PCB and TCDD concentrations. The subcohort in the ten Tusscher study was small, and therefore the results are not as robust as those in the Weisglas-Kuperus et al. (2004) study, which included almost 91% of the original 2000 study cohort.

Animal studies also suggest that the developing immune system is sensitive to persistent changes in immune function or status, especially when exposure occurs in the perinatal and neonatal stage and especially in T-cell-mediated immunity. Less compelling studies exist from which to estimate the sensitivity of the adult human immune system.

EPA concludes that

> there is insufficient clinical data from these studies to fully assess human sensitivity to TCDD exposure. Nevertheless, based on the results of the extensive animal work, the database is sufficient to indicate that the immune effects could occur in the human population from exposure to TCDD and related compounds at some dose level. At present, it is EPA's scientific judgment that TCDD and related compounds should be regarded as nonspecific immunosuppressants and immunotoxicants until better data to inform judgment are available. (Reassessment, Part III, p. 2-37, lines 4 to 10)

Indeed, based on the extensive animal data, it is reasonable and prudent for EPA to regard TCDD as an immunotoxicant. Furthermore, the Dutch study provides some suggestive evidence for this conclusion. However, it is unclear in EPA's conclusion what is meant by TCDD and related compounds being regarded as "nonspecific immunosuppressants and immunotoxicants" and this should be clarified.

How Persistent Are the Immunotoxic Effects of TCDD on the Human Immune System (Reversible Versus Irreversible Changes)?

Because it has not yet been unequivocally established that TCDD induces immune suppression in humans, it is not possible at this time to delineate the persistence of TCDD-mediated immunotoxicity in humans. The lowered total white-blood-cell numbers reported in the studies of Dutch children (Ilsen et al. 1996) were no longer evident 2 years after birth (Weisglas-Kuperus et al. 2000). The elevated T-cell subpopulations in Dutch children at 42 months of age did not appear to persist at 8 to 10 years of age (ten Tusscher et al. 2003). However, Weisglas-Kuperus et al. (2004) also reported a positive association in children (3 to 7 years of age) between increased postnatal PCB exposure and increased prevalence in recurrent middle ear infections. In addition, there was a positive association between increased prenatal PCB exposure and decreased chicken pox frequency as well as allergy and asthma. A second recent report (Van den Heuvel et al.

2002) suggested that TCDD, other dioxins, and DLCs may produce subtle but persistent changes in immune status as evidenced by a reduction in allergy and asthma. In this study of 200 Flemish adolescents, a negative correlation was observed between exposure to TCDD, other dioxins, and DLCs with TCDD TEQ measured and allergic responses in airways. In addition, serum immunoglobulin G levels were also negatively correlated with PCB exposure.

Relevance of Rodent Models to Human Quantitative Risk Assessment for Immunotoxicity

In the Reassessment, Part III, Appendix A, Table A-1, EPA centers its risk characterization for adult immunological effects on four studies conducted in mice (Vecchi et al. 1983; Narasimhan et al. 1994; Smialowicz et al. 1994; Burleson et al. 1996) and its developmental immunotoxicological risk characterization on a single rat study (Gehrs and Smialowicz 1999) (see Appendix A, Table A-1; for immune end points; see Figures 12 to 15, pp. A-7 to A-9; same studies identified in Table 5-6, p. 5-39, in the Reassessment, Part III). Based on these studies, lowest-observed-adverse-effect levels (LOAELs) and no-observed-adverse-effect levels (NOAELs) are used to derive ED_{01} (1% effective dose) and human equivalent intake values. Because of the importance that these studies have to the Reassessment and the potential importance that the derived values may have for risk management, some additional comments are provided here.

The study of Burleson et al. (1996) showed the lowest LOAEL (6 ng/kg), and NOAEL (3 ng/kg) values, which result in calculated "human equivalent intakes" of 1 pg/kg/day and 2 pg/kg/day, respectively. The study showed that the sensitivity to alteration by TCDD of host resistance of mice to H3N2 influenza A (Hong Kong/8/68) virus is strikingly sensitive compared with other LOAELs for immunotoxicological end points in adult rodents given in Table A-1, the LOAEL ranging from 100 to 1,200 ng/kg. In fact, the sensitivity to TCDD in the Burleson et al. study is striking even when compared with similar and more recently published host-resistance studies using influenza virus. For example, Nohara et al. (2002) showed that TCDD doses up to 500 ng/kg did not increase mortality in a number of different strains of mice, including B6C3F1, C57Bl/6, Balb/c, and DBA/2 mice infected with influenza A virus (A/PR/34/8, H1N1). More mice per group were used in the Nohara study than in the Burleson study, thus providing even greater statistical power. In a study using influenza A/HKx31, Warren et al. (2000) reported that in certain experiments, TCDD treatment (1 to 10 μg/kg) increased mortality, whereas in other experiments no mortality was observed. Furthermore, Warren and coworkers stated that in some experiments TCDD doses as high as 10 μg/kg produced no mortality, whereas 80% mortality was observed at the same dose in other experiments. Such data emphasize the variability typically observed in host-resistance studies. The reason for the significantly greater sensitivity to TCDD in the Burleson et al. (1996) study is unclear but strongly suggests that further studies are needed before using results from this study for risk characterization.

The LOAEL, NOAEL, ED_{01}, and human equivalent intake values for immunotoxicity are based on suppression by TCDD of the anti-SRBC IgM antibody-forming cell response in studies by Vecchi et al. (1983), Narasimhan et al. (1994), and Smialowicz et al. (1994) (the other three studies identified in Table A-1 and used for risk characterization of the adult immune system). Numerous laboratories have demonstrated suppression of the antibody-forming cell response by TCDD, and in general, there is good concordance in the ED_{50} doses (600 to 770 ng/kg) derived from these studies (see Table 4-1, p. 4-38) (Vecchi et al. 1980; Davis and Safe 1988; Kerkvliet and Brauner 1990; Kerkvliet et al. 1990). In contrast, some variability in the LOAEL values identified in Table A-1 were observed in three other studies: 100 ng/kg (Narasimhan et al. 1994), 300 ng/kg (Smialowicz et al. 1994), and 1,200 ng/kg (Vecchi et al. 1983). The variation is due primarily to dose selection in each of the studies. It is clear how the LOAEL and, in the case of the Narasimhan study, the NOAEL were identified from the data presented in each of the published reports. It is not clear, however, how the ED_{01} values were calculated.

Several concerns also exist about using the Gehrs and Smialowicz (1999) study for characterizing developmental immunotoxicological risk by TCDD. The Gehrs and Smialowicz study gives no indication as to the number of rat offspring studied; therefore, it is unclear whether the results are robust. In addition, both males and females were found to be more sensitive to immune suppression by TCDD after 14 months of age than at 4 months of age, as measured by a delayed type hypersensitivity response, which is somewhat puzzling. Although hypotheses could be advanced to explain these unexpected findings, it would be valuable and prudent to repeat that study before using those results for characterizing developmental immunotoxicological risk by TCDD.

Congruence with Full Document

In Part II, Chapter 4 of the Reassessment, a comprehensive and balanced synthesis of the immunotoxicology literature on TCDD, other dioxins, and DLCs is presented. Results from more than 200 published studies are discussed in an organized and logical manner. Moreover, there is good congruence between Chapter 4 and section 2.2.3 on immunotoxicity in the Executive Summary.

CONCLUSIONS AND RECOMMENDATIONS ON THE IMMUNOTOXICITY OF TCDD, OTHER DIOXINS, AND DLCS

• Present clinical findings are inconclusive about whether or in what way TCDD, other dioxins, and DLCs are immunotoxic in humans, a conclusion that EPA acknowledges, and human data are also sparse. Perhaps the most compelling data that these compounds are human immunotoxicants, at possibly relevant environmental levels, come from the studies of the Dutch children's cohort. These studies show an association between prenatal exposure and changes in immune status. However, the effects are modest and do not lie significantly outside the full range of normal. The correlation of increased otitis media in the very young with perinatal TEQ is the only statistically significant immunological clinical finding. Some of the same findings were made in acutely exposed Taiwanese and Japanese cohorts (Yu et al. 1995). Concordant with findings in Dutch children are a number of animal studies that also suggest that the developing immune system is especially sensitive to modulation by TCDD, other dioxins, and DLCs. Collectively, in light of the large database showing that these compounds are immunotoxic in laboratory animal studies together with sparse human data, EPA is being prudent in judging TCDD, other dioxins, and DLCs to be potential human immunotoxicants in the absence of more definitive human data.

• EPA's conclusion that TCDD, other dioxins, and DLCs are immunotoxic at "some dose level" by itself is inadequate. At a minimum, a section or paragraph should be added that discusses the immunotoxicology of these compounds in the context of current AHR biology. Specifically, there is evidence showing that the affinity of TCDD for the human AHR is at least an order of magnitude lower than that in high-responding $Ahr^{b\text{-}1}$ mouse strains (Ramadoss and Perdew 2004), which has been the most commonly used animal model for investigations of immunotoxicity of TCDD, other dioxins, and DLCs. Other properties of AHR, in addition to binding affinity, such as specificity for target genes and transactivation potential, will contribute to the toxicity produced by AHR ligands. Nevertheless, EPA supports a TEF/TEQ approach for estimating the immunotoxic potency of mixtures of dioxins, other than TCDD, and DLCs. The Reassessment assumes that immunotoxicity is therefore primarily mediated through an AHR-dependent mechanism, so some discussion should be included acknowledging the possibility that rodents, especially certain mouse strains expressing $Ahr^{b\text{-}1}$ might be significantly more sensitive to the immunotoxic effects of TCDD, other dioxins, and DLCs than humans. Some discussion should also be included on the strengths and weaknesses of using genetically homogeneous inbred mice to characterize immunotoxicological risk in the genetically variable human population. Expanding the discussion to include the above crucial points would provide additional balance to Part III of the Reassessment.

- EPA centers its risk characterization for adult immunological effects on four studies conducted in mice (Burleson et al. 1996; Smialowicz et al. 1994; Narasimhan et al. 1994; Vecchi et al. 1983), and its developmental immunotoxicological risk characterization on a single rat study (Gehrs and Smialowicz 1999) (see Part III, Appendix A, Table A-1; for immune end points; see Figures 12 to 15, pp. A-7 to A-9). Concerns about Table A-1 are the following:
- The calculations of ED_{01} values and the scientific assumptions made in deriving those values need further clarification. Likewise, EPA should provide a clear scientific rationale for selecting ED_{01} as a benchmark dose.
- Considerations of the Burleson et al. (1996) study with no consideration of two similar studies—Nohara et al. (2002) and Warren et al. (2000)—that yield very different results requires justification.
- On the basis of concerns discussed earlier, it would be prudent to replicate the Gehrs and Smialowicz (1999) study before using its results for characterizing developmental immunotoxicological risk of TCDD.

An important animal study by Oughton et al. (1995) was not included in either Part II or the tables in the Executive Summary of the Reassessment. The importance of the study is that it is the only low-level chronic exposure investigation published (TCDD at 200 ng/kg/week once a week to Bb mice at 2 to 16 months of age) in which immunotoxicological parameters have been assessed—specifically, a phenotypic analysis by flow cytometry of major cell subpopulations in the mouse spleen, thymus, and peripheral blood. The study showed only subtle alterations in the immune system as demonstrated by a modest increase in γ-δ T cells, which the authors considered "questionable biological relevance," and a small decrease in the frequency of memory CD4 cells (by phenotype). However, these changes, although of questionable biological relevance, have also been observed in humans and in high-exposure animal studies.

REPRODUCTION AND DEVELOPMENT

Animal Data

EPA provided an overview of the effects of TCDD, other dioxins, and DLCs on development and reproduction based on published animal studies and accidental human exposures. Determination of the alterations in development and reproduction is a highly complex process because hormonal as well as intracellular processes and compensatory mechanisms, including hormonal feedback mechanisms, are affected. The Reassessment comprehensively covers developmental aspects in a wide variety of models. Two major rodent models have been used to study the effects of TCDD on reproduction and development. In the first model, TCDD is given during pregnancy (an in utero and lactational exposure model). This model tested the ability of TCDD to disrupt development of the pups and assessed the effects on reproduction and reproductive behavior later in life. A comprehensive overview of the in utero and lactational exposure model is presented, but the doses used and how the model relates to the human reproductive and developmental toxicity are not emphasized. For example, maternal concentrations of TCDD in plasma are needed at designated times during pregnancy and lactation in the rat dam; those data would allow comparison to human data and determination of whether concentrations in rodents are higher or lower than those in humans accidentally exposed to TCDD. In the second model, adult rats and an immature gonadotropin-primed model are used to assess the effects of TCDD on ovulation. These models were not adequately discussed. In addition, there is uncertainty in the risk assessment based on differences in TEF reported by the World Health Organization (WHO) (Van den Berg et al. 1998) when compared with published data. For example, the WHO 1998 TEQs appear to be 2.5-fold higher (Van den Berg et al. 1998) than actual potency data determined with these models (Safe 1990; Gao et al. 1999, 2000a,b). Studies have been conducted using the 1998 TEFs (e.g., Hamm et al. 2003), and the conclusions seemed to indicate that mixture doses two to three times higher than the calculated TEQ appeared to be required to elicit the same alterations. The study by Hamm et al. is comprehensive and revealed numerous adverse effects on male and female reproduction, such as prolonged time to puberty, decreased seminal vesicle and ventral prostate weights, and, in the female, increased the incidence of vaginal threads. The lowered responses to the mixture of TCDD, other dioxins, furans, and coplanar PCBs were attributed to decreased transfer of mixture components to the offspring, whereas a miscalculation of the TEQ might have also contributed to the lowered response (Hamm et al. 2003). In addition, if the mixture was altered to favor what might have been present in the diet in nature, then the true TEQ might not have been accurately represented in the Hamm et al. study. However, given that the WHO TEFs are order-of-magnitude estimates of the relative potency of a chemical and derived from all toxicological outcomes in a variety of species, it is not surprising that there is a lack of absolute concordance between a calculated TEQ and an actual TEQ, measured in one species, for male and female reproduction end points. The generation of end-point-specific TEFs would probably resolve the observed differences between calculated and measured TEQs on these end points.

Examples of complete dose responses on various reproductive parameters in females using various polychlorinated dibenzo-*p*-dioxins (PCDDs), polychlorinated dibenzofurans (PCDFs), and PCBs are given below. Trace levels of PCDDs have been detected in fish, wildlife, and humans (Van den Berg et al. 1985; Tiernan et al. 1985), and PCDDs have toxicological effects on the reproduction and development in vertebrates, including rodents and nonhuman primates (Van den Berg et al. 1994). Few studies have evaluated the effects of complex mixtures of PCDDs and DLCs on the female reproductive system. Previous studies have validated the TEQ concept for several PCDDs in acute and subchronic/chronic experiments using several biological end points (Stahl et al. 1992, Weber et al. 1992, 1994; Rozman et al. 1993, 1995; Viluksela et al. 1997a,b and 1998a,b).

Female reproductive toxicity of TCDD is evidenced by reduced ovulation (Li et al. 1995a,b) and developmental defects (Heimler et al. 1998a), which were orders of magnitude less than the cancer response (Kociba et al. 1978; 1979; Rozman et al. 1993), indicating that ovulation and development are more sensitive end points because lower doses are needed to disrupt reproductive processes than to increase the incidence of cancer. Studies using gonadotropin-treated immature rats revealed that complex mixtures of PCDDs, such as TCDD, pentachlorodibenzo-*p*-dioxin (PeCDD), and hexachlorodibenzo-*p*-dioxin (HxCDD), as well as each congener alone produced dose responses that lowered ovarian weights and the number of ova shed (Gao et al. 1999). In addition, the effects of PCDDs were additive when an equipotent mixture of the PCDDs was given. The slopes of the dose-response curves were not statistically different among the various congeners. Thus, the additive effect and parallel dose-response curves indicated a similar mechanism of action. The PCDFs and PCBs, with TCDD-like actions, also have a similar inhibitory effect on ovulation. The studies of Gao et al. (1999) and others (Krishnan and Safe 1993) are in close agreement and indicate a TEF of 0.12 to 0.2 for PeCDD, which differs from the TEF of 0.5 proposed by WHO (Van den Berg et al. 1998). The doses required in the ovulation study for PCDDs (Gao et al. 1999) to produce the same effect increased approximately 5-fold for each chlorine added to TCDD. Those observations are consistent with prior studies (Stahl et al. 1992) and imply a TEF based on female reproductive effects of 0.2 for PeCDD and 0.04 for HxCDD, which differ from the WHO report in which a TEF of 1 was given for PeCDD (Van den Berg et al. 1998) and used by Hamm et al. (2003) for numerous reproductive studies in males and females.

TEFs for the pentachloro-isomers of PCDPs and PCBs are in the same range as those previously reported for other end points (Safe 1990; Van Birgelen et al. 1994a,b, 1996). However, the WHO conference (Van den Berg et al. 1998) reported values of 0.5 and 0.1 for 2,3,4,7,8-pentachlorodibenzofuran (PeCDF) and 3,3′,4,4′,5-PCB, respectively, which are twice as high as most studies report. The doses of the pentachloro-isomers of PCDs and PCBs studied by Gao et al. (2000b) had 10-fold lower potency than the ED_{50} for TCDD in blocking ovulation (8 μg/kg) (Gao et al. 2000b). Generally, TEF values are combined from all end points, and development of end-point-specific TEFs might ultimately be useful. However, it must be noted that TEFs are not expected to be exact, and the values determine in the reproductive studies are well within an order of magnitude.

The Reassessment was mainly directed at understanding the adverse effects of TCDD administered during pregnancy on development of the pups; that section was superbly written and covered numerous aspects of exposure to TCDD in utero and during lactation. However, little risk estimate information is given. Also, an important part of the literature on the adult female reproductive system was not addressed in the Reassessment. This included mechanisms of ovulatory blockage at the level of the hypothalamic-pituitary axis and the ovary and endocrine disruption of reproductive processes by TCDD in adult rodents (Goldman et al. 2000; Petroff et al. 2001; Valdez and Petroff 2004). The committee summarizes some of those studies in the following section.

Effects of PCDDs on the Ovary

Studies have shown that PCDDs adversely affect ovarian function by direct actions on the ovary and the hypothalamic-pituitary axis (Petroff et al. 2000; Valdez and Petroff 2004). Human ovarian follicular fluid has been found to contain PCDDs (Tsutsumi et al. 1998), implicating PCDDs in possible adverse ovarian effects. Exposure of adult female rats to PCDDs disrupted estrous cycles, delayed ovulation, and lowered ovarian weights (Li et al. 1995a; Cummings et al. 1996). Irregular menstrual cycles were observed in female rhesus monkeys fed TCDD in the diet (Allen et al. 1977; Barsotti et al. 1979). Mice are less prone to the adverse ovarian effects of PCDDs in some studies (Cummings et al. 1996), although TCDD caused the formation of ovarian cysts in CD-1 mice (Gallo et al. 1986). In rats, administration of TCDD before mating interrupts fertility by affecting both ovulation and implantation (Giavini et al. 1983). In the immature gonadotropin-primed rat, the adverse effects of PCDDs on the ovary were characterized by small ovaries, the absence of corpora lutea, and numerous unruptured preovulatory follicles (Gao et al. 1999, 2000b; Petroff et al. 2001). In the immature rat primed with gonadotropin, the number of ova shed in response to PCDDs was dose-dependently inhibited with an ED_{50} of TCDD at 8 μg/kg of body weight. This supported the TEQ for several other AHR agonists, including PeCDD, HxCDD, PeCDF, and pentachlorobiphenyl (PeCB) (Li et al. 1995a,b; Gao et al. 1999, 2000b; Son et al. 1999; Petroff et al. 2000, 2001). TCDD suppressed follicular development as determined by a reduction in the number of antral and preantral follicles in the pups of pregnant rats exposed to TCDD in utero and during lactation (Heimler et al. 1998a). The anovulatory effect of PCDDs was observed in gonadotropin-primed hypophysectomized rats (Gao et al. 1999; Petroff et al. 2000; Roby 2001), and direct application of TCDD to the ovary blocked ovulation as well (Petroff et al. 2000). Thus, PCDDs have direct effects on the ovulatory follicle that are sufficient to block ovulation.

The rat ovary expressed AHR mRNA (Son et al. 1999), and macaque granulosa cells also expressed AHR mRNA which was increased by human chorionic gonadotropin (Chaffin et al. 1999). β-Naphthoflavone (Bhattacharyya et al. 1995) and TCDD (Son et al. 1999) also increased ovarian Cytochrome P4501A1 protein (CYP1A1) mRNA in rats. The direct effects of PCDDs on ovarian steroid production are less clear, despite consistent blockade of ovulation after systemic and local ovarian exposure to PCDDs. In immature gonadotropin-primed female rats, pretreatment with PCDDs increased serum estradiol during the preovulatory period and reduced serum progesterone concentrations consistent with blockage of ovulation and reduced luteinization (Gao et al. 1999, 2000b). In addition, in the immature rat model, PCDDs blocked the surges of follicle-stimulating hormone (FSH) and lutenizing hormone (LH) in sera on expected proestrus (Li et al. 1995b; Gao et al. 1999, 2000b). Collectively, these results indicate that the adverse effects of PCDDs may be due to effects on gonadotropin release as well as to direct effects on the ovary (Son et al. 1999; Petroff et al. 2000; Roby 2001). In CD-1 mice and avian species, PCDDs did not alter serum concentrations of estradiol (DeVito et al. 1992; Janz and Bellward, 1996).

In vitro models have been used to assess the effects of PCDDs on ovarian steroidogenesis. PCDDs decreased cellular uptake of glucose and reduced protein kinase A activity and secretion of progesterone and estradiol in human granulosa cells (Enan et al. 1996a,b). However, another study reported an initial inhibition of estradiol in human luteinized granulosa cells that was followed by increased estradiol accumulation at 36 and 48 hours (Heimler et al. 1998b). A decrease in aromatase activity and reduced messenger ribonucleic acids (mRNAs) for P450ssc and P450arom in FSH-stimulated rat granulosa cells exposed to PCDDs has been reported (Dasmahapatra et al. 2000). In contrast, PCDDs failed to alter progesterone, androstenedione, or estradiol secretion in in vitro cultures of whole ovarian dispersates, granulosa cells, or thecal-interstitial cells derived from immature rats (Son et al. 1999) Although, this lack of in vitro action is also seen in immune cells in vitro.

One target of PCDDs may be alterations in follicular proteolysis and tissue remodeling during the periovulatory period as ovulation is blocked after acute exposure to TCDD, other dioxins, and DLCs (Gao et al. 1999, 2000b; Petroff et al. 2001). Potential mechanisms blocking degradation of the follicular wall may involve modulation of steroid action since decreased expression of ovarian cyclooxygenase-2 (COX-2) and AHR coincide with increased plasminogen activator inhibitor-1 (PAI-1) and tissue plasminogen activator (PA) (Mizuyachi et al. 2002). Because PA participates in ovulation in the rat (Tsafriri 1995), TCDDs may increase PAI-1, reduce overall PA activity, and block ovulation. It is well known that PA activity increases after the ovulatory surges of LH and FSH as a result of increased granulosal cell prostaglandin secretion, a process dependent on COX-2 (Richards et al. 1987). COX-2 has TCDD response elements. Thus, TCDD, other dioxins, and DLCs may block ovulation by inhibiting granulosal prostaglandin secretion, reducing COX-2 in the preovulatory follicle, before reducing PA activity after an increase in ovarian PAI-1.

TCDD reduces expression of the progesterone receptor (PR), and PR null mice do not ovulate (Lyndon et al. 1996). TCDD is well known to inhibit estradiol-induced PR in the breast cancer cell line MCF-7 through an AHR-mediated mechanism (Harper et al. 1994). However, within 24 hours after administration of TCDD to immature rats, estrogen receptor (ER)α, ERβ, and PR were unaffected in the ovaries (Mizuyachi et al. 2002). Thus, the role of the PR in the antiovulatory effects of PCDDs is unresolved.

Effects of PCDDs on the Hypothalamus and Pituitary Gland

TCDD, other dioxins, and DLCs reduce pituitary secretion of LH and FSH at the time of the LH and FSH surges but premature surges of LH and FSH have been reported in immature rats (Gao et al. 1999; 2000a,b). In the Han Wistar rat that is resistant to TCDD, (50 μg/kg) caused atrophy of the pituitary with little to no loss of weight and no mortality (Pohjanvirta et al. 1993). However, exposure of fetal (in utero) or neonatal (via mother's milk) mice to TCDD reduced pituitary weights of male offspring (Theobald and Peterson 1997).

LH synthesis in the pituitary is controlled by gonadotropin-releasing hormone (GnRH) and gonadal steroids feed back negatively to reduce secretion. LH and FSH secretion were altered in gonadotropin-primed female rats pretreated with TCDD, other dioxins, and DLCs (Li et al. 1995b; Gao et al. 1999, 2000a,b). TCDD-treated animals had reduced gonadotropin secretion during the preovulatory period compared with controls. Culture of pituitary halves with TCDD dose-dependently reduced LH secretion, but no effect of TCDD was observed in primary pituitary cell cultures (Li et al. 1997).

Preovulatory increases in estradiol are required through a positive-feedback mechanism for induction of the LH and FSH surges on proestrus. TCDD has antiestrogenic effects and inhibits ovulation through blockage of the LH and FSH surges. However, serum concentrations of estradiol in intact control and TCDD-treated rats are similar during the preovulatory period, indicating the possibility that the lack of estradiol action was causal in blocking the surges (Li et al. 1995b; Gao et al. 1999, 2000a,b). This appeared to be the case, as a long-acting exogenous estradiol administered during TCDD treatment overcame the blockade on ovulation and restored the LH and FSH surges (Gao et al. 2001).

Exogenous GnRH also overcame the inhibitory effects of TCDD on ovulation by restoring the LH and FSH surges in the immature gonadotropin rat model (Gao et al. 2000a). Controls exhibited normal LH and FSH surges, whereas such surges diminished in rats treated with TCDD. GnRH treatment increased secretion of LH and FSH to surge levels in TCDD-treated rats and partially restored ovulation. Those data indicate that GnRH secretion may have been reduced by TCDD. The failure of the gonadotropin surges to completely restore ovulation in rats receiving TCDD and GnRH indicates the possibility that adverse direct effects of TCDD on the ovaries may have reduced the number of ovulations.

Effects of TCDD on the Cardiovascular and Pulmonary Systems

Since the publication of EPA's draft Reassessment, a substantial body of literature has emerged concerning the effects of TCDD on heart and vascular development. The developing vascular system appears to be a target very sensitive to TCDD in vertebrate embryos. Much of the work in this area has been performed in zebrafish (*Danio rerio*) embryos, a model that has the advantage over mammalian and avian models of allowing for direct visual observation of many developing organ systems, including the heart and associated vasculature. Studies have also been performed in avian and rodent models.

Several studies have indicated a fundamental role for the AHR system in vascular development and hence a theoretical basis for the TCDD sensitivity of vascular development. Lahvis et al. (2000) generated *Ahr*$^{-/-}$ mice that displayed reduced liver size. Developing mice exhibited altered vascular architecture, including massive portosystemic shunting due to a patent ductus venosus, resulting in reduced blood flow to the liver and hence reduced hepatocyte size and liver mass. This failure of the ductus venosus to close in *Ahr*$^{-/-}$ mice was subsequently associated with major hepatic veins failing to decrease in size, as observed in wild-type mice, which may result in increased blood pressure or a failure in vasoconstriction (Lahvis et al. 2005). Walisser et al. (2004b) observed that mice engineered to contain a hypomorphic *Arnt* allele (underexpressing ARNT, the AHR nuclear translocator protein) demonstrated the same vascular phenotype and were resistant to TCDD toxicity versus wild-type mice. Together with the AHR studies, this indicated essential roles for ARNT and for AHR-ARNT dimerization for both the purported developmental and TCDD toxicity roles of the AHR pathway. TCDD exposure during a specific time frame of embryonic development rescued vascular development in both *Ahr* and *Arnt* hypomorphs, indicating the requirement for activation of the AHR-ARNT heterodimer for normal vascular development (Walisser et al. 2004a).

Studies with fish models, particularly zebrafish, have demonstrated the sensitivity of the cardiovascular system, including cardiomyocytes, to TCDD during embryonic development (Antkiewicz et al. 2005). Studies with morpho-lino antisense oligonucleotides to knock down expression of specific genes in the zebrafish embryo have supported the key role of AHR in the developmental effects of TCDD. Knockdowns of AHR2 prevented TCDD-induced pericardial edema, trunk circulation failure, and anemia in developing zebrafish (Prasch et al. 2003; Dong et al. 2004). (Due to gene duplication, zebrafish have two AHRs, AHR1 and AHR2; TCDD-mediated effects are associated with binding to AHR2, and not to AHR1.) In these studies, the AHR2 morpholinos were highly effective at blocking TCDD-induced cytochrome P4501A protein (CYP1A) expression in the vascular endothelium. Carney et al. (2004) showed that, whereas an AHR2 morpholino protected zebrafish from TCDD-mediated effects of reduced blood flow to trunk segments and pericardial edema, the CYP1A morphlino did not provide protection against TCDD toxicity in contrast to the findings of Teraoka et al. (2003). Collectively, these studies demonstrate that these developmental effects of TCDD are AHR2 mediated in zebrafish, but the role of CYP1A remains unresolved. TCDD has also been demonstrated to perturb cardiovascular development in the chicken embryo (Sommer et al. 2005) and in maternally exposed fetal mice (Thackaberry et al. 2005a,b). In addition, cardiovascular function is compromised in *Ahr*$^{-/-}$ mice (Lund et al. 2005; Vasquez et al. 2003).

These studies addressing the effects of TCDD on cardiovascular development were not performed with the objectives of quantitative risk assessment in mind. However, given the sensitivity of this end point at a very sensitive lifestage, EPA is encouraged to consider these and related studies identifying adverse effects of TCDD on cardiovascular development and function in its risk assessment for noncancer end points.

Human Data

The Reassessment extensively documents the known reproductive, developmental, and ectodermal consequences of TCDD exposure in a variety of animal species (Part II, Chapter 5) and describes to a lesser extent various other noncancer consequences, including hepatic, thyroid, and cardiovascular effects observed in animals other than humans (Part II, Chapter 7, part B). In assessing the potential for related risks in humans, EPA makes several critical assumptions.

Assumption: Because dioxins are proven causes of reproductive, developmental, and other abnormalities in various animal species, they may, therefore, cause similar effects in humans. (Part III, p. 2-33, lines 3 to 5; p. 6-1, lines 21 to 22; p. 6-3, lines 14 to 16).

For reproductive, developmental, and ectodermal effects, this assumption is readily justified given the nature and extent of the animal data. Further, the profiles of reported human reproductive, developmental, and ectodermal effects after exposures to TCDD, other dioxins, and DLCs are similar to the effects found in animals, thus lending overall general support to the assumption. Similarities in developmental effects are most compelling at the highest levels of exposure such as those reported in the Yusho and Yu-Cheng poisonings (Part II, pp. 5-15 to 5-16) because "all four manifestations of developmental toxicity (reduced viability, structural alterations, growth retardation, and functional alterations) have been observed to some degree" (Part II, p. 5-97, lines 1 to 3).

Even so, the developmental effects are not entirely consistent and the Reassessment appropriately notes that other than the mouse "no other species develops cleft palate except at maternal doses that are fetotoxic and maternally toxic" (Part II, p. 5-19, lines 10 to 11) and that "studies in humans have not clearly identified an association between TCDD exposure and structural malformations" (Part II, p. 5-19, lines 15 to 17). As discussed below, the effects of low-level TCDD exposure on reported human developmental effects are less compelling. Although the spectrum of reported human reproductive and hormonal abnormalities following TCDD exposure is generally similar to that found in animals, the strengths of the individual associations in studies thus far, are weak, and confidence in the causal nature of these associations while suggestive is not yet compelling.

In reference to other noncancer consequences of TCDD exposure, the assumption remains equally valid, although the animal evidence for other noncancer end points, such as adverse effects on hepatic enzymes (Part II, section 7.15.1.2.3), pancreatic islet function (Part II, section 7.15.2.1.2), thyroid hormone dysregulation (Part II, section 5.2.3.6; Part III, section 2.2.1.3), lipid abnormalities (Part III, section 2.2.6.3), and cardiopulmonary or circulatory disturbances (Part II, section 7.15.3.1; Part III, section 2.2.6.3), is often more limited in scope.

Assumption: Humans are neither more nor less sensitive than animals as far as the adverse effects of dioxins are concerned (Part II, p. 8-4, lines 6 to 27; Part III, p. 2-3, lines 28 and 29; p. 2-32, lines 14 to 16]. Given the paucity of systematic in vivo human data, this assumption is the parsimonious choice and also the most defensible based on in vitro data (Part II, pp. 8-4 to 8-5). Nevertheless, EPA acknowledges the uncertainty and imprecision of this assumption noting that (1) "for most toxic effects produced by dioxin, there is marked species variation" (Part II, p. 8-5, line 31); (2)

human epidemiological studies are confounded by the fact that the unexposed "cohorts contain measurable amounts of background exposure to PCDDs, PCDFs, and dioxin-like PCBs" (Part II, p. 8-5, line 35; p. 8-6, line 1); (3) "many epidemiological studies are hampered by small sample size, and in many cases the actual amounts of TCDD and related compounds in human tissues were not examined" (Part II, p. 8-6, lines 2 and 3); (4) "it is often difficult, if not impossible, to assess in humans the same endpoints that might be determined in experimental animals" (Part II, p. 8-6, lines 4 and 5); and (5) "it is essentially impossible to determine the contribution of TCCD-like versus non-TCDD-like congeners to fetal/neonatal toxicity" (Part II, p. 5-15, lines 14 and 15) in the poisoning episodes where complex mixtures containing a variety of toxicants were ingested accidentally (Part III, p. 2-23, lines 32 to 35).

Assumption: Noncancer effects can occur at body burden levels in animals equal to or less than body burdens calculated for tumor induction in animals (Part III, p. 5-25, lines 28 and 29). Although critical to the discussion of noncancer end points in humans, the strength of this assumption is unknown and the uncertainty is possibly large. The propagated uncertainties leading to this assumption are highly dependent on the inherent uncertainties in the use of TEQs, the calculation of the historical body burden, and the modeling of dose-response effects, as discussed in detail in Chapters 2 and 3. Because of limited epidemiological evidence, further uncertainty is introduced by the inability to demonstrate convincing associations and dose-response relationships between TCDD exposure and noncancer end points in humans (Part III, p. 2-23, lines 20 to 22), as discussed below.

Assumption: ED_{01} is an acceptable departure point for calculating the risks of noncancer end points. As noted above, the limitations of this assumption are highly dependent on the inherent uncertainties in the use of TEQs, the calculation of body burden, and the modeling of dose-response effects, as discussed in detail in Chapters 3 and 5.

The EPA Reassessment does not adequately discuss the level of confidence that should be accorded results whose statistical significance is associated with wide uncertainty limits. Attention should also be directed to addressing the potential biological significance of very small statistically significant physiological or biochemical changes that remain well within the normal range of variation and adaptation.

Furthermore, the EPA Reassessment continues to rely on the approach that diverse human data collected across disparate studies of different types and inherent strengths can be interpreted with confidence without applying the more formalized tools of evidence-based medicine. Thus, the EPA Reassessment (as well as Institute of Medicine [IOM] committee report) relies largely on committee-based, consensus evaluation of the available data rather than on specifically commissioned, rigorous analyses constructed according to established criteria that both formally evaluate the strengths of the available evidence and integrate, by quantitative systematic review, the data across available studies (Sackett et al. 2000; NCI 2002; CEBM 2005; Guzelian et al. 2005).

On the whole, the potential for increased risk of noncancer end points after exposure to TCDD at or near background levels is cautiously presented in the Reassessment. However, the Reassessment explicitly characterizes TCDD as "developmental, reproductive, immunological, endocrinological, and carcinogenic hazards" (Part III, p. 6-3, lines 10 and 11), although the formal criteria for defining human hazard in the context of these noncancer end points are not defined precisely in the Reassessment. Further, although the Reassessment acknowledges that "some have argued that in the absence of better human data, deducing that a spectrum of noncancer effects will occur in humans overstates the science" (Part III, p. 6-3, lines 33 and 34), the EPA position is that an inference of human effects "is reasonable given the weight of evidence from available data" (Part III, p. 6-4, lines 1 and 2). Nonetheless, as EPA concedes, available human data currently do not permit resolution of these divergent evaluations.

Human Reproductive and Developmental Outcomes

The available human reproductive and developmental studies available at the time of the Reassessment draft are presented in detail, although a number of the more recent follow-up studies are obviously not reported, as mentioned below. EPA provides an overall conclusion that "subtle effects, such as the impacts on ... developmental outcomes ... or the changes in circulating reproductive hormones in men exposed to TCDD, illustrate the types of responses that support the finding of subtle yet arguably adverse effects at or near background body burdens" (Part III, p. 6-2, lines 6 to 11). The committee agrees that the results are subtle but disagrees that the reported effects are truly clinically adverse, especially when confidence in the observations is low and the reported changes could be non-significant at the biological level and clinical outcome. In this context, the Reassessment also notes that "there is no reason to expect, in general, that humans would not be similarly affected [as animals] at some dose, and a growing body of data supports this assumption. On the basis of the animal data, current margins of exposure are lower than generally considered acceptable, especially for more highly exposed human populations. The human database supporting this concern for potential effects near background body burdens is less certain" (Part III, p. 6-32, lines 19 to 23).

Male Reproductive Hormones

The Reassessment's description of the National Institute of Occupational Safety and Health (NIOSH) study report (Egeland et al. 1994) showing a significant positive correlation of serum LH and FSH levels with serum TCDD does not discuss the weak nature of this correlation, the wide confidence intervals (CIs) around the regression, or the hormone values within the normal range (Part II, section 7.13.5.1). Similarly, the text further describes a two to four times higher prevalence of low testosterone levels among workers exposed to TCDD but does not report that the CIs around the risk ratios at the higher serum TCDD levels not only are very broad but also cross 1.0, indicating limited confidence in the significance of the relationships (Part II, section 7.13.5.1). Nor does the EPA Reassessment report that no dose-response effect was observed (odds ratio = 3.9 at lowest range of TCCD levels and 2.1 at highest levels), although the 95% CIs of the odds ratios themselves are so broad as to raise significant uncertainty about whether there is indeed a dose response relationship indicated by these studies (Part II, section 7.13.5.1).

Similarly, the Reassessment states that the Ranch Hand study (Roegner et al. 1991) (Part II, section 7.13.5.1) reported lower serum testosterone levels in Ranch Hand veterans with current serum TCDD levels exceeding 33.3 pg/g, although the reported difference (10.2 ng/dL) was "statistically nonsignificant" and unlikely to have a measurable physiological effect. The EPA Reassessment also describes three additional negative studies (CDC 1988; Grubbs et al. 1995; Henriksen et al. 1997), concluding that "the human data offer some evidence of alterations in male reproductive hormone levels associated with substantial occupational exposure to 2,3,7,8-TCDD" (Part II, p. 7B-38). Thus, although "some evidence" has been reported, the bulk of the reported evidence is either negative or uncertain to a degree.

The Department of Defense (DOD) released the latest report of the Ranch Hand study in 2005. The committee did not have the opportunity to review the report in detail because its release coincided with the end of the committee's deliberations. However, the document reports that "the difference in adjusted free testosterone means in Ranch Hand versus Comparisons was 10.95 versus 10.47, respectively. The LH means for Ranch Hand and Comparison officers were 4.49 mIU/mL versus 4.09 mIU/mL, respectively. Both were well within one standard deviation of normal-age matched populations. No evidence of a dose-response effect was seen based on categorized dioxin or 1987 dioxin levels" (DOD 2005, p. 18-156). The report concludes that "the association of dioxin with ... gonadal abnormalities appeared weak at best and unlikely to be clinically significant" (DOD 2005, p. 18-156) and "associations between dioxin level and ...

gonadal hormone abnormalities were unlikely to be clinically important" at these levels (DOD 2005, p. 21-8).

Female Reproductive System

In the Reassessment's discussion of potential effects of TCDD exposure on endometriosis, the more recent Seveso data (Eskenazi et al. 2002a) are not included. Compared with women with TCDD concentrations of ≤20 ppt, the relative risk of endometriosis is 2.1 in women with TCDD concentrations >100 ppt, but the 90% CI ranges from 0.5 to 8.0, indicating little confidence in the true magnitude of the rate ratio or the significance of the reported average relative risk of 2.1. One conclusion from these data might be that women whose serum TCDD levels were >20 ppt had no more endometriosis than those whose serum TCDD concentrations were ≤20 ppt. Another defensible conclusion might be that the study did not have the power to come to any convincing conclusion on this issue. The authors of the study, Eskenazi et al. (2002a), chose to describe their findings as a "doubled, nonsignificant risk for endometriosis among women with serum TCDD levels of 100 ppt or higher, but no clear dose response." Finally, in a recent review of the nonhuman primate and the human data assessing the relationship between TCDD exposure and endometriosis Guo concluded that "there are no solid, credible data available at this moment to support the hypothesis that dioxin exposure may lead to the development of endometriosis" (Guo 2004).

Data published within the past 2 years on effects of exposure to TCDD, other dioxins, and DLCs on the menstrual cycle in women are obviously not referenced in the 2000 Reassessment. Thus, data from the Seveso incident surveying women who were exposed to TCDD postnatally, but while they were prepubertal, found "no change in the risk of onset of menarche with a 10-fold increase in TCDD," and there was "no evidence of a dose-response trend" (Warner et al. 2004). Likewise, postmenarchal women exposed in Seveso showed no association of TCDD exposure with menstrual cycle length, but, in women exposed before menarche who had a 10-fold increase in serum TCDD concentrations, the menstrual cycle was lengthened by 0.93 day, although the 95% CI ranged from –0.01 to 1.86, and the strength of the relationship between menstrual cycle length and serum TCDD concentration shown in Figure 1A of the report is not convincing, with widely scattered data points (Eskenazi et al. 2002b). An observational study of wives and sisters of Swedish fishermen found a 0.49-day shorter menstrual cycle (95% CI 0.03 to 0.89) in those with a high dietary exposure to polychlorinated organochlorine compounds, including TCDD, but found no association with early life exposure (Axmon et al. 2004).

The discussion on spontaneous abortions briefly mentions the study on the NIOSH cohort as "in press" (Part II, section 7.15.3.4.5). This study has now appeared (Schnorr et al. 2001) and found no effect on the incidence of spontaneous abortion or on the sex ratio of offspring. The authors concluded that the study provided "additional evidence that paternal TCDD exposure does not increase the risk of spontaneous abortions at levels above those observed in the general population." Likewise, recent data from the Seveso cohort (Eskenazi et al. 2003) showed no association of TCDD with spontaneous abortions.

Other recent relevant studies include birth-weight results reported for the NIOSH cohort (Lawson et al. 2004) and the Seveso cohort (Eskenazi et. al. 2003). The recent NIOSH report (Lawson et al. 2004) found that paternal TCDD exposure had no effect on birth weight for term infants, and a "somewhat protective" association of preterm delivery with paternal TCDD (odds ratio = 0.8), although the 95% CI ranged from 0.6 to 1.1. There was no obvious increase in birth defects, although the results were descriptive only. The authors concluded that "because the estimated TCDD concentrations in this population were much higher than in other studies, the results indicate that TCDD is unlikely to increase the risk of low birth weight or preterm delivery through a paternal mechanism." The recent Seveso follow-up (Eskenazi et al. 2003) also showed no association of TCDD concentration with offspring birth weight or with the birth of infants small for gestational age. Finally, the Reassessment does not discuss the female Vietnam veterans study reported by Kang et al. (2000), which also reported no increase in spontaneous abortions, stillbirths, low-birth-weight infants, or infant deaths among women veterans who had served in Vietnam (and possibly exposed) compared with those who had served in the United States, although there are no body burden measurements made.

The data on birth-weight effects were described adequately (Part II, section 7.13.12.8), and the summary comment (Part II, section 7.13.12.9) reflected appropriately the uncertainty of whether there were any birth-weight effects of exposure to TCDD at the time of the Reassessment. However, the likelihood of TCDD exposure having a measurable effect on birth weight has been substantially reduced by the recently reported studies of Kang et al. (2000), Eskenazi et al. (2003), and Lawson et al. (2004). To reflect and appropriately weigh this new information, EPA should correspondingly modify the summary comment (Part II, section 7.13.12.9).

For the state of available information in 2000, the Reassessment describes adequately the observed effects of TCDD exposure on offspring sex ratio (Part II, sections 7.13.12.7, 7.15.3.4.8). As noted in the report, increased female births were observed after the Seveso accident (Mocarelli et al. 1996, 2000). They were also observed in a study of offspring of Russian pesticide producers exposed to TCDD (Ryan et al. 2002). However, the

Schnorr study mentioned in the Reassessment as "in press" has now been published (Schnorr et al. 2001) and found no effect of TCDD exposure on sex ratio of offspring in the NIOSH cohort.

In the United States, both exposure to TCDD and the male-to-female sex ratio at birth have declined since the early 1970s (Matthews and Hamilton 2005). This parallel decline is opposite that postulated from TCDD poisoning incidents. Thus, because sex ratios at birth not only undergo temporal trends but also show racial and nationality differences and are affected by both maternal age and infant birth order (Matthews and Hamilton 2005), EPA should also acknowledge the uncertainties inherent in evaluating sex ratios at birth without properly controlling for the aforementioned variables. The committee recognizes, however, that the TCDD exposure studies that showed altered gender ratios at birth have reported ratio values that were greater than the changes that might normally be expected to be caused by the variables mentioned above.

Childhood Growth and Postnatal Development

The Reassessment text describes appropriately the cited growth data and conveys adequately the uncertainty of whether TCDD exposure has effects on postnatal growth in humans (Part III, section 7.13.12.9). The issue would not be further clarified by including the omitted Swedish fish exposure study (Rylander et al. 1995) that reported diminished height, but not weight, at age 18, because this report includes neither TCDD nor TEQ data. Two of the longest-term studies of chlorinated toxicant effects on growth published subsequently (Blanck et al. 2002; Gladen et al. 2000) deal with PCB exposure and thus do not contribute to resolving the debate about the effects of TCDD on childhood growth.

Similarly, the longest neurodevelopmental follow-up studies (Jacobson and Jacobson 1996; Gray et al. 2005) are reports on PCB exposure and do not directly contribute to the current TCDD issues since no TEQ is derived. However, the ongoing Dutch follow-up study referenced repeatedly in the Reassessment has now published its findings in 6.5-year-old children (Vreugdenhil et al. 2002a). At that age, there were no cognitive or motor differences between breast-fed infants (primarily postnatally exposed) and formula-fed infants (primarily exposed in utero with background postnatal exposure), including no overall differences in global cognitive index, memory, or motor performance, except when children from "less optimal homes" were analyzed separately. This observation suggested to the authors that less optimal home environment may allow the effects of TCDD on neurodevelopment to become manifest more readily while, in more optimal home environments, the additional beneficial environmental influences overcome the detrimental effects of exposure to TCDD. This is clearly a hypothesis at this stage, given the availability of only this single study that has addressed the issues. In an additional report, Vreugdenhill et al. (2002b) also described decreased masculinized play in boys and increased masculinized play in girls at age 7.5 years. Although statistically significant, the biological relevance of these conclusions remains uncertain given the wide scatter of the data and the regression coefficients reported.

Cardiovascular and Pulmonary Systems

The Reassessment discusses in detail the available data on potential human cardiopulmonary consequences of TCDD exposure highlighting the difficulty of supporting firm conclusions about the presence of a relationship (Part II, sections 7.13.9, 7.13.9.1, 7.13.10). Recently, from the latest data on Ranch Hands (DOD 2005), DOD concluded that "no consistent evidence suggested that herbicides or dioxin were associated with ill effects on respiratory health" (p. 21-9). On the other hand, "the presence of heart disease was found to be higher among Ranch Hands than Comparisons in enlisted flyers" (p. 21-6), and "an increased percentage of Ranch Hands in the high dioxin category were found to have abnormally high diastolic blood pressure. Ranch Hands in both the low dioxin category and the low and high dioxin categories combined were found to have a lower mean systolic blood pressure. Similarly, a smaller percentage of Ranch Hands in both the low dioxin category and the low and high dioxin categories combined had an abnormally high systolic blood pressure" (p. 21-6). However, the report notes that "the prevalence of cardiovascular disease was not increased in the Ranch Hand cohort. In only one analysis, that of diastolic blood pressure noted above, was there any evidence of an increased risk with increased body burden of dioxin" (p. 21-7).

OTHER NONCANCER END POINTS

Diabetes

The Reassessment (Part II, sections 7.13.6, 7.13.6.1, 7.15.2.1.2; Part III, section 2.2.5) presents in detail the then available data on the relationship between TCDD exposure and the development of Type 2 diabetes. This relationship was evaluated in greater depth by an IOM committee (IOM 2000), which concluded that "there is limited/suggestive evidence of an association between exposure to the herbicides used in Vietnam or the contaminant dioxin and Type 2 diabetes." This is an adequate statement of the state of the science concerning this noncancer end point, and the committee recommends that EPA revise the Reassessment to include the analysis provided by the IOM committee. The Reassessment should also incorporate the data of the recent Ranch Hand study report (DOD 2005) showing that "mean fasting insulin and the risk of diabetes requiring insulin control increased with initial dioxin. C-peptide and time to diabetes onset decreased as initial dioxin increased." The risk of diabetes requiring insulin control was increased in the Ranch Hand high-dioxin category. An increase in the risk of diabetes requiring oral hypoglycemic or insulin control was observed as 1987 dioxin levels increased. Time to diabetes onset decreased as 1987 dioxin levels increased. The risk of an abnormally high hemoglobin A1c increased with 1987 dioxin levels. Some findings in the DOD (2005) report appeared inconsistent with the results presented above, such as a decrease in the risk of 2-hour postprandial urinary glucose abnormalities with 1987 dioxin levels. The findings appear consistent with the previously noted association between Type 2 diabetes and dioxin in Ranch Hand veterans. Increased risks of diabetes requiring insulin control were found with initial dioxin, in the high-dioxin category, and with 1987 dioxin levels. In contrast, "associations between dioxin level and thyroid and gonadal hormone abnormalities were unlikely to be clinically important" (DOD 2005, p. 21-8).

These data led to the conclusion that "the association noted at previous Air Force Health Study examinations between Type 2 diabetes mellitus and dioxin persisted. A higher prevalence of diabetes, as well as severity, as dioxin increased was evident, even after adjustment for such factors as age and body mass index" (DOD 2005, p. 18-156).

Thyroid Function

The Reassessment acknowledges that despite the fact that "many effects of TCDD exposure in animals resemble signs of thyroid dysfunction or significant alterations of thyroid-related hormones" (Part III, p. 2-41, lines 32 and 33), the results of human studies "are mostly equivocal" (Part III, p. 2-42, line 1). The Reassessment reports that Pavuk et al. (2003) showed "elevated TSH [thyroid stimulating hormone] means among the high TCDD exposure group in the 1985 and 1987 follow-ups, with an increasing trend across the decade 1982-1992, but no association with the occurrence of thyroid disease" (Part III, p. 2-42, lines 2 to 4). The discussion does not address the fact that the TSH differences, although statistically significant, are quantitatively extremely small and well within the normal range of circulating TSH levels. Further, limitations of the study are not described, including 86 exclusions "from the longitudinal analyses because they had undergone thyroidectomy, or had endocrine cancer, or were on thyroid medication" (Pavuk et al. 2003), the exclusion of one Ranch Hand because of an extremely low TSH value that might have indicated the earliest sign of hyperthyroidism and the fact that "over the five examinations, three different radioimmunoassays were used to measure TSH" (Pavuk et al. 2003). In addition the Reassessment does not report that no significant differences "with regard to mean levels of total thyroxine (T4), triiodothyronine (T3)% uptake, or free thyroxine index were observed at any examination" (Pavuk et al. 2003). Recently, DOD (2005) reaffirmed that "as for a dioxin effect related to thyroid disease, the 2002 examination data did not support such a relation."

Further, while the most recently published Ranch Hand follow-up report (DOD 2005) found a difference in adjusted mean TSH levels (1.653 microunits/mL versus 1.557 microunits/mL) in Ranch Hands and Comparisons, respectively, this difference was "not considered clinically significant because a 1% difference is difficult to measure. The same was true from the free T4 values in enlisted flyers (mean of 1.115 ng/dL in Ranch Hands versus 1.054 ng/dL in the comparison groups). If a primary thyroid effect were present, one would expect the TSH to move in the opposite direction of the free T4, which was not seen in these data" (DOD 2005, p. 18-156).

The draft Reassessment also highlights the higher TSH values reported in human infants by Pluim et al. (1993) and by Koopman-Esseboom et al. (1994) (Part III, 2.2.6.2 and Part II,7.15.2.2.2), but does not discuss the fact that the TSH changes were very small and possibly not of physiological or clinical significance. Follow-up of the Dutch children's cohort has now been carried out for more than a decade, and changes in thyroid status in this cohort have not been reported, although it is unclear whether they were in fact thoroughly assessed. The text in the Reassessment does include lengthy hypothetical discussions (Part III, 2.2.1.3 and Part III, 2.2.6.2) of plausible mechanisms for perturbations of human thyroid function, although there is limited human data to support or refute such mechanisms.

Teeth

Because ectodermal abnormalities are common findings in animal studies, including nonhuman primates, as well as in the human Yusho and Yu-Cheng exposures, the enamel hypomineralization found in 6- to 7-year old Finnish children (Alaluusua et al. 1996, 1999) is highlighted twice (Part II, section 7.13.12.6.1; Part III, section 2.2.2.1). Because the enamel mineralization scores were largely subjective, it is imperative that the observers were blinded to the prior breast-feeding status of the children, but this issue is not specifically mentioned in the publications. Additionally, the Reassessment highlights several other limitations of this study and, in the related commentary (Part II, section 7.13.12.6.2) EPA acknowledges that "the presentation of the results is incomplete."

Chloracne

The Reassessment adequately summarizes the uniformly agreed upon and well-documented association between TCDD exposure and the development of chloracne.

Elevated γ-Glutamyl Transferase

The Reassessment adequately summarizes the data showing "a consistent pattern of increased γ-glutamyl transferase (GGT) levels among individuals exposed to TCDD-contaminated chemical." (Part III, p. 2-40, lines 23 and 24) and notes that "long-term pathological consequences of elevated GGT have not been illustrated by excess mortality from liver disorders or cancer; or in excess morbidity in the available cross-sectional studies" (Part III, p. 2-41, lines 23 and 24). The report also acknowledges that "the consistency of the findings in a number of studies suggests that the finding may reflect a true effect of exposure but for which the clinical significance is unclear" (Part II, p. 7B-116, lines 20 to 22).

The most recent report of the Ranch Hand study found no relationship between TCDD exposure and GGT (DOD 2005, pp. 13-51 to 13-57).

Lipid Levels

The Reassessment accurately notes that "neither adults nor children from Seveso had lipid levels above the referent level" (Part II, p. 7B-118) despite very high exposures and that "the most recent data suggest that high exposure to 2,3,7,8-TCDD contaminated substances are not related significantly to increased lipid concentrations, specifically total cholesterol and triglycerides" (Part II, p. 7B-118). The Reassessment adds that "slight but chronic elevations in serum lipids may put an individual at increased risk for disorders such as atherosclerosis and other conditions affecting the vascular system" (Part II, p. 7B-118) and that "risk factors such as dietary fat intake, familial hypercholesterolemia, alcohol consumption, and exercise" (Part II, p. 7B-118) were not considered in the Seveso study, even though no effects of TCDD exposure were found.

The most recent Ranch Hand follow-up showed no relationship between TCDD exposures and total high-density-lipoprotein cholesterol, although ranch hands had an increased percentage of individuals with increased triglyceride values (DOD 2005, pp. 13-78 to 13-103). This report concluded that "based on the analysis of triglycerides, a subtle relation between dioxin and lipid metabolism cannot be excluded" (DOD 2005, p. 13-218).

The various remaining noncancer end points, for which there are fewer data and even less suggestive evidence than the associations discussed above, are adequately summarized throughout the Reassessment.

CONCLUSIONS AND RECOMMENDATIONS ON THE REPRODUCTIVE, DEVELOPMENTAL, AND OTHER NONCANCER END POINTS OF TCDD, OTHER DIOXINS, AND DLCS

- Embryonic and fetal development and female and male reproduction are sensitive end points of toxicity from TCDD, other dioxins, and DLCs in rodents because, as discussed earlier, responses occur at lower administered doses than other end points. However, the sensitivity of these end points in humans is less apparent.
- The fetal rodent is more sensitive than the adult rodent to adverse effects of TCDD.
- The human equivalent intake (pg/kg-day) for some adverse effects related to reproduction and development based on ED_{01} is not adequately supported (Chapters 2, 3, and 4).
- In humans, there is a clear association of TCDD exposure with chloracne and available studies have shown suggestive associations of TCDD exposure with Type 2 diabetes, but the latter data are not yet robust.
- In humans, the association of TCDD exposure with other reported, detrimental noncancer effects has not been convincingly demonstrated. The available studies have not yet shown clear associations among TCDD exposures and the risks of individual, clinically significant, noncancer end points.
- In reference to human disease risks, the overall conclusions about noncancer risks due to TCDD exposure are, in general, cautiously stated, and the uncertainty of suspected relationships is acknowledged. Nonetheless, the limitations of specific human studies are not uniformly addressed, and the broad 95% CIs accompanying some reported statistically significant effects are not discussed in the context of the uncertainty that these broad confidence limits imply. Conversely, statistically insignificant effects are sometimes highlighted.
- The divergent data across the diverse studies assessing human noncancer end points have not been subjected to systematic review according to currently accepted approaches, including meta-analysis when appropriate, nor has there been formal grading of the quality of the evidence according to accepted principles currently applied in other areas of clinical pathophysiology, including one report of the relationship of TCDD exposure to cancer end points (Crump et al. 2003).
- EPA should discuss how the ED used in the in utero and lactational exposure rat models relates to human reproductive and developmental toxicity and risk information, including TEFs and TEQs.

• EPA should more clearly describe how and why ED_{01} values were determined in animals and transferred to human equivalents for the various noncancer end points and address risk estimate calculations using alternative assumptions (e.g., ED_{05}). Whereas ED_{01} is conceptually a viable POD, the committee has concerns about how the ED_{01} is computed and whether there are adequate data at the ED_{01} level to ensure an acceptable level of confidence in the conclusions derived from using the ED_{01}. The dynamic range approach EPA used to compute ED_{01} for continuous response is flawed in that the change of 1% total range may not identify any meaningful toxic effects, that 1% change may be well within random variation in the absence of exposure, and that the use of total range is less sensitive than use of a control range because total range can be much wider.

• EPA should provide a discussion of the dose-response effects of TCDD, other dioxins, and DLCs on the adult female reproductive system that result in endocrine disruption in animals. The impact of the dose-response data provided in these studies on human risk assessment should be presented.

• With respect to human noncancer end points, the Reassessment text should be revised to include the relevant, more recent data and, when appropriate as discussed above, to reflect study quality and data uncertainty of the studies referenced.

• For available human, clinical, noncancer end point data, EPA should establish formal principles of and a formal mechanism for evidence-based classification and systematic statistical review, including meta-analysis when possible. The application of systematic review, followed by evidence-based classification, leads to a more explicit statement of, and concrete appreciation of, the level of certainty (and, correspondingly, of uncertainty) that can be accorded the answers to specific questions in a particular field.

• When the mechanism is established, currently available and newly available human clinical studies should be subject to such systematic review and formal evidence-based assessment. The quality of the available evidence should be reported, and the strength or weakness of a presumptive association should be classified according to currently accepted criteria for levels of evidence. Animal studies have shown that TCDD can cause a variety of noncancer effects. These studies support both the EPA position of the plausibility of corresponding human effects and the need to devise adequately designed investigations that will answer the questions in man.

• In making its final recommendations, EPA should incorporate and integrate the relevant data from both human and animal studies, as appropriate, according to the levels-of-evidence hierarchy devised.

• EPA is encouraged to review newly available studies on the effects of TCDD on cardiovascular development in its risk assessment for noncancer end points.

7

Review of Risk Characterization

The Reassessment[1] (Part III, Chapter 6) considers risk characterization under a series of headings, many of which represent summaries of the inputs to risk characterization instead of the output of risk characterization and its formulation of advice to risk managers. For convenience, this chapter uses the same headings for the committee's review of the Reassessment's risk characterizations before presenting its conclusions. Because Chapter 6 in the Reassessment summarizes data from previous chapters, many of the committee's comments here were raised in previous chapters of this report.

REVIEW

TCDD, Other Dioxins, and DLCs Can Produce a Wide Variety of Effects in Animals and May Produce Many of the Same Effects in Humans

Reassessment (Part III, pp. 6-1 to 6-4)

This introductory text sets the scene by stating:

> Effects will likely range from detection of biochemical changes at or near background levels of exposure to detection of adverse effects with increasing severity as the body burdens increase above background levels. (Part III, p. 6-1, lines 28-30)

---

[1] *The Exposure and Human Health Reassessment of 2,3,7,8-Tetrachlorodibenzo-p-Dioxin (TCDD) and Related Compounds* (EPA 2003a, Part I; 2003b, Part II; 2003c, Part III) is collectively referred to as the Reassessment.

> Clearly adverse effects, including, perhaps, cancer, may not be detectable until exposures contribute to body burdens that exceed current backgrounds by one or two orders of magnitude (10 to 100 times). (Part III, p. 6-2, lines 11-13)

The rationale for those statements is not clearly defined, although the Reassessment states later that few clinically significant effects were detected in the small number of human cohorts studied, nearly all members of which had body burdens significantly above background levels.

The text considers species differences in sensitivity to 2,3,7,8-tetrachlorodibenzo-*p*-dioxin (TCDD, also referred to as dioxin), other dioxins, and dioxin-like compounds (DLCs) and concludes that "humans most likely fall in the middle rather than at either extreme of the range of sensitivity for individual effects among animals" (p. 6-2, lines 25-26). A general comparison across several species is not relevant to the focused issue of risk characterization. The comparisons of importance are those between humans and the species and strains used in the specific studies that revealed adverse effects at the lowest levels of exposure, the so-called critical effects (see below). This general statement on interspecies differences detracts from a critical and quantitative assessment of differences in sensitivity between humans and the species used in the key toxicological studies for risk characterization.

Overall the committee considered this introductory section to be reasonable but unfocused.

TCDD, Other Dioxins, and DLCs are Structurally Related and Elicit Their Effects Through a Common Mode of Action

Reassessment (Part III, pp. 6-4 to 6-5)

The text is uncontroversial and concludes that binding to the aromatic hydrocarbon receptor (AHR) appears to be necessary but is not sufficient to elicit the various TCDD-induced effects. The committee agrees with this conclusion.

EPA and the International Scientific Community Have Adopted Toxic Equivalency of TCDD, Other Dioxins, and DLCs as a Prudent Scientific Policy

Reassessment (Part III, pp. 6-5 to 6-6)

The text summarizes the current situation and is uncontroversial. Obviously, given the date of the Reassessment, the text has not considered planned updates to toxic equivalency factor (TEF) values and whether these will be used in future risk assessments of TCDD, other dioxins and DLCs. The committee's recommendations related to TEFs appear in Chapter 3.

Complex Mixtures of TCDD, Other Dioxins, and DLCs Are Highly Potent, Likely To Be Carcinogens

Reassessment (Part III, pp. 6-6 to 6-12)

The Reassessment states that "because TCDD, other dioxins, and DLCs always occur in the environment and in humans as complex mixtures of individual congeners, it is appropriate that the characterization [likely carcinogen] apply to the mixture" (p. 6-6, lines 19-21). Therefore, despite the attention given by EPA and hence by this committee to consideration of whether TCDD is "carcinogenic to humans" or "likely to be carcinogenic to humans," the reality is that TCDD is always present as part of a mixture of TCDD, dioxins, and DLCs, and, therefore, the practical hazard characterization of human exposure to TCDD is in effect considered by EPA "likely to be carcinogenic." In consequence, the focus on qualitative classification of the nature of the cancer hazard by EPA has been a somewhat futile exercise. The text then further discusses this issue and concludes that the "likely to be carcinogenic" classification could differ in strength, depending on the constituents in the mixture. Subsequent text reconfirms that TCDD is classified by EPA as "carcinogenic to humans" and outlines the evidence used to reach this conclusion, including the presence of "strong and consistent" evidence from occupational epidemiological studies (a point on which the committee does not agree; see Chapter 5). The committee concluded that such detailed consideration of hazard classification was of little value in a section on risk characterization, especially as the difference between "carcinogenic" and "likely to be carcinogenic" would not have a significant impact on the formulation of advice under risk characterization.

This section continues by presenting the upper bound of the cancer risk estimate of 1×10^{-3} per pg of toxic equivalent quotient (TEQ)/kg of body weight/day for both background intakes and incremental intakes above background. The value is based on the range of cancer slope factors (CSFs) developed from linear modeling of the occupational cohort data. The Reassessment states, "Evaluations of shape parameters...for biochemical effects that can be hypothesized as key events in a generalized dioxin mode-of-action model do not argue for significant departures from linearity below a calculated ED_{01}, extending down to at least one to two orders of magnitude lower exposure" (p. 6-8, lines 31-33, to 6-9, lines 1-2). This sentence appears to be critical to the risk characterization approach for cancer adopted by EPA, but there is no scientific assessment of the strength of the available evidence to support that statement, of the ability of the model-fitting methods used by EPA to detect a departure from linearity, were one to exist; or of the indications of nonlinearity in the dose-response relationships for many noncancer effects also considered to be mediated via an interaction with the AHR. The Reassessment attempts to explain the decision to use a linear approach (Part III, p. 5-15, lines 27-29) by stating that "The linear default is selected on the basis of the agent's mode of action when the linear model cannot be rejected and there is insufficient evidence to support an assumption of nonlinearity." Quantitative evidence of nonlinearity below the point of departure (POD), the ED_{01} (effective dose), will never be available because the POD is chosen to be at the bottom end of the available dose-response data. As discussed in Chapter 5 of this report, EPA should give greater weight to knowledge about the mode of action and its impact on the shape of the dose-response relationship. The committee considers that the absence of evidence that argues against linearity is not sufficient justification for adopting linear extrapolation, even over a dose range of one to two orders of magnitude or to the assumption of linearity through zero, which would not normally be applied to receptor-mediated effects. This view is supported by the results of the recent cancer bioassay (NTP 2004), which was not available to EPA at the time of the Reassessment but which could have a major impact on the risk assessment approach adopted by EPA.

The text compares the current estimate with previous EPA estimates and uses previous evaluations with the same approach to support the outcome of the Reassessment. The difference between EPA's evaluation and that of the Food and Agricultural Organization of the United Nations (FAO)/World Health Organization (WHO) Joint Expert Committee on Food Additives (JEFCA) (which considered TCDD to be a nongenotoxic carcinogen and an uncertainty factor approach to be adequate to account for both cancer risk and noncancer effects) was thought to "reflect differences in science policy" (p. 6-11, line 2). The Reassessment did not attempt to explain why EPA has chosen to use an uncertainty factor approach for the risk characterization of other nongenotoxic, receptor-mediated carcinogens with a known mode of action (such as for thyroid carcinogens) but not for TCDD. The Reassessment suggests that a margin-of-exposure (MOE) approach should be adopted for both cancer and noncancer effects but does not explore the implications of the estimated MOE for cancer or the ability of the MOE approach to refine the advice for population groups.

The use of different methods for the risk characterization of end points that result from the same basic underlying mode of action is scientifically illogical, a conclusion that seems to be supported by EPA in an earlier part of the Reassessment (Part III, p. 5-3, lines 18 to 28).

Although the Reassessment defines a slope factor and cancer risk estimate, it does not spell out clearly the health implications of emphasizing this approach for the U.S. population. The Reassessment states that the slope factor has a "public health-conservative nature" (p. 6-9, line 14), but such risk management considerations should not be used to support an approach to risk characterization or detract from selection of the most appropriate, scientifically justifiable approach.

The text discusses the consistency of the present slope factor with previous EPA evaluations and further discusses the issue of hazard classification, as if the decision to define a CSF affected the hazard classification (or vice versa), which illustrates the lack of clarity and focus in this part of the Reassessment.

The Reassessment recognizes that "the shape of the dose-response curve below the range of observation can be inferred only with uncertainty" (p. 6-12, lines 1 to 2), and therefore the Reassessment should have given equal weight and critical evaluation to the derivation of a CSF and to the calculation of the MOE with a discussion of the adequacy of the MOE of an exposure in relation to the remaining uncertainties.

In future revisions of the Reassessment, aspects that should be discussed for both approaches include the known mode of action, the adequacy of the occupational cohorts to represent the whole population, the integration of data from the animal cancer bioassays (including the most recent study) in relation to the spectrum of cancers detected, and the shape of the dose-response relationship.

Use of a MOE Approach to Evaluate Risk for Noncancer and Cancer End Points

Reassessment (Part III, pp. 6-12 to 6-18)

Despite the title of this section, there is no focused discussion in the Reassessment of the MOE in relation to cancer or to each of the end points, using exposure data relevant to that end point (Part III, Appendix A, Table A-1). In addition, there is no discussion of the areas of uncertainty that would need to be taken into account for each study and end point. For example, the MOE for cancer, and possibly for immune and neurodevelopmental effects, would be based on epidemiological data, whereas MOEs for noncancer effects would be based largely on data from animal studies. Other issues that should be discussed in the interpretation of the MOE for each end point are the relevance of the effect to the general population and to population groups and life stages and, most important, the clinical significance of the magnitude of the effect detected at the ED_{01} (if this is retained as a point of comparison on the dose-response relationship; see committee comments earlier in this report).

The Reassessment concludes that setting a reference dose (RfD) is not appropriate because of the relatively high background levels of exposure compared with effect levels (Reassessment, Part III, p. 6-14) and that "any RfD that the Agency would recommend using a traditional approach for setting an RfD using uncertainty factors to account for limitations of knowledge is likely to be below—perhaps significantly below (by a factor of 10 or more)—current background intakes and body burdens." EPA thus concluded that setting an RfD would be of little value for evaluating possible risk management options when the RfD has already been exceeded by average background exposure. This issue is not resolved by simply replacing the RfD by an MOE without analyzing whether the estimated MOE is adequate for that particular end point based on the data used to derive the point of comparison on the dose-response relationship.

The magnitude of this apparent problem arises from two aspects of the Reassessment: the use of ED_{01} as the point of comparison with exposure for continuous variables, which in many cases is two orders of magnitude below the lowest-observed-adverse-effect level (LOAEL) (Reassessment, Appendix A, Table A-1), and the use of the usual default uncertainty factors despite the wealth of data available on TCDD. The issue of the derivation and suitability of the ED_{01} for continuous variables was discussed in Chapter 2 of this report. EPA has not justified its use for risk assessment and its replacement of traditional measures, such as the no-observed-adverse-effect level (NOAEL), LOAEL, or BMD_{10} (benchmark dose) or BMD_{05}, as a point for establishing an RfD or for comparison with human exposures by calculation of MOEs. Selection of appropriate uncertainty factors is discussed further below.

An additional problem identified by the Reassessment is that "the calculation of an RfD (with its traditional focus on a single "critical" effect) distracts from the large array of effects associated with similar body burdens of dioxin" (Part III, p. 6-14, lines 20-22). This statement appears to contradict EPA's well-established approach of focusing on the critical effect as a basis for setting health protective values. The problem applies to some degree to many other chemicals that show multiple effects over a narrow dose range and does not invalidate the selection of a "critical effect" from the TCDD database, which would lead to a more focused discussion and risk characterization. A critical effect of postnatal reproductive changes after in utero exposure was identified from the available dose-response data by the European Scientific Committee on Food (SCF 2000, 2001) and by JECFA (2002). These bodies concluded that developing a health-based guidance value (equivalent to an RfD) from reproductive effects in rats after in utero exposure would also cover the risks of other effects (including cancer) detected at slightly higher body burdens. In principle, the case of TCDD is no different from that of other contaminants that produce multiple adverse effects. Because the exposures of a proportion of the U.S. population would be above any RfD, it would have been useful for EPA to define the nature and magnitude of the risks at different levels of intake, the groups of the population most at risk, and the major sources of exposure for any at-risk groups. Alternatively, if MOEs were calculated for noncancer effects, then the risk characterization should describe the nature of the adverse effect and the uncertainties and variability inherent in both the BMD (ED) estimate and the relevant exposure estimate. It would have been useful if MOE values had been calculated and discussed for different exposure scenarios.

The Reassessment discusses the approaches adopted by the SCF and WHO (IPCS1998b) and by the Agency for Toxic Substances and Disease Registry (ATSDR 1998) (Reassessment, Part III, pp. 6-16 to 6-18). The more recent JECFA (2002) evaluation was considered with reference to the date of the meeting in 2001. The Reassessment highlights three sources of differences:

1. An initial focus on cancer or noncancer effects.
2. The use of intake or body burden.
3. The "safety" or "uncertainty" factors used.

The recent evaluations by SCF and JECFA used the approach proposed by the WHO International Programme on Chemical Safety (IPCS) in which the usual 10-fold default uncertainty factors are subdivided into toxicokinetic and toxicodynamic subfactors, which can be replaced when suitable chemical-specific data are available (IPCS 1994, 1999, 2004; WHO 2005). In this approach, the 10-fold interspecies factor is divided into 4.0 for toxicokinetics and 2.5 for toxicodynamics, the product (10) being used in the absence of chemical-specific data to replace either of the default values; the 10-fold human variability factor is subdivided equally into 3.2 for toxicokinetics and 3.2 for toxicodynamics. Subdividing the 10-fold default uncertainty factors was done by EPA in its recent evaluation of boron (EPA 2004b), although a slightly different split of the 10-fold interspecies factor was used. (This difference between EPA and WHO would not have altered significantly the health-based guidance value for TCDD, other dioxins, and DLCs that was derived by SCF and JECFA.)

The Reassessment considers briefly the rationale for the uncertainty factors used in the recent SCF and JECFA evaluations. These evaluations concluded that interspecies differences in toxicokinetics had been taken into account by the use of body burden as the dose metric instead of the external dose, and therefore this subfactor would become 1 instead of 4.0. The Reassessment does not discuss the explanation of why SCF and JECFA concluded that "no uncertainty factor needed to be applied for differences in toxicodynamics between experimental animals and humans and for interindividual variation [in toxicodynamics] among humans" (Reassessment, Part III, p. 6-17, lines 21-23). The rationale given in the JECFA monograph (JECFA 2002, p. 590) was that, in general, rats are more sensitive than humans to the adverse effects of TCDD, and therefore the interspecies factor might be less than 1, but that "it cannot be excluded that the most sensitive humans might be as sensitive to the adverse effects of TCDD as rats were in the pivotal studies. Therefore, it was concluded [by the JEFCA] that no safety factor in either direction need to be applied for differences in toxicodynamics among humans." In other words, any possible variability in toxicodynamics among humans would be compensated for by the higher inherent sensitivity of the rat strains used in the pivotal studies compared with average humans, and each of these subfactors would become 1. Of the four aspects for which the usual 100-fold uncertainty factor is applied, the only one for which data was considered to be inadequate related to human variability in toxicokinetics. SCF and JECFA applied the default value of 3.2 for human variability in toxicokinetics.

While the approach adopted by the SCF and JECFA is open to criticism because of its simplicity, the attempt to incorporate the wealth of data on TCDD into the risk assessment process contrasts with EPA's assumption that default values would be used, and hence the RfD would be below the current levels of exposure. The Reassessment (Part III, p. 6-18, lines 6-9) states, "In particular, the focus on accounting for residual toxicodynamic differences in cross-species scaling and interindividual variability in the general population to account for sensitive individuals, including children" would suggest larger uncertainty factors than have been proposed by these groups if EPA were to set an RfD. However, EPA does not discuss how the usual uncertainty factors might be modified using the TCDD database and does not give an analysis of the uncertainty factors that it would use and justification for their use. The Reassessment does not discuss whether or not the EPA considered how the uncertainty factors or other aspects of risk characterization could be revised based on probabilistic approaches.

The Reassessment does not evaluate critically the extent of species differences in target organ sensitivity, especially in relation to the pivotal studies and critical effects. Overall, there is inadequate discussion of the relative affinities of the AHRs in rats and humans and of the possible impact of polymorphisms in *AHR* and other sources of sensitivity differences within humans. The Reassessment (Part III, p. 6-18, line 4-5) states, "Traditional approaches that might be applied by EPA or that have been applied by ATSDR would likely require additional information to support the choice or removal of uncertainty factors as performed by WHO, SCF and JECFA." However, there is no critical discussion of the limitations of the available data that might be used to move away from the traditional uncertainty factors. The Reassessment does not give the rationale for EPA's decision not to replace the default uncertainty factors by chemical-specific data, despite the enormity of the TCDD database.

The Reassessment (Part III, p. 6-18, lines 10-14) concludes that "any composite uncertainty factor greater than 10 applied to effect levels based on body burden ... would result in TDI or MRLs below the current background intakes. The use of uncertainty factors in the range of 30 to 100 or more, as traditionally used by EPA, would result in values even further below some current background body burdens."

The Reassessment concludes the risk characterization section (Part III, p. 6-34) by stating that the MOEs based on body burden are less than 1 for enzyme induction in rats and mice and less than 4 for developmental effects in rats and endometriosis in nonhuman primates. The reader is left to compare those values with uncertainty factors in the range of 30 to 100, which EPA would traditionally use, with no clear and concise guidance on the interpretation of this information. However, these judgments are based on the nontraditional use of ED_{01} in place of a BMD_5, BMD_{10}, NOAEL, or LOAEL.

Children's Risk from Exposure to TCDD, Other Dioxins, and DLCs May Be Increased, but More Data Are Needed to Address This Issue

Reassessment (Part III, pp. 6-18 to 6-21)

The Reassessment highlights the greater susceptibility of in utero, perinatal, and neonatal life stages on the basis of animal and human epidemiological data. The Reassessment does not clarify the additional data that would be required before an RfD could be established or before definitive advice could be given about the adequacy or inadequacy of the MOE for adverse effects detected in animal studies after in utero exposure. Following these general doubts about the possible heightened susceptibility of neonates and children, the Reassessment comments on the greater exposure of nursing infants and children but concludes that, because the risk characterization is based on body burden, the short-term intake levels will have little impact on risk compared with overall lifetime exposure. The committee noted that EPA did not define the MOE for these life stages and that, overall, this section raises concerns about hypothetical, additional, undefined susceptibility while allaying concerns about the considerably greater exposures of infants and children compared with adults.

Background Exposures to TCDD, Other Dioxins, and DLCs Need To Be Considered When Evaluating Hazard and Risk

Reassessment (Part III, pp. 6-21 to 6-23)

This section of the Reassessment provides a summary of the extent of background exposures but does not adequately integrate the information into an MOE or an estimate of population cancer risk using the slope factor.

Evaluating the Exposure of "Special" Populations and Developmental Stages Is Critical to Risk Characterization

Reassessment (Part III, pp. 6-23 to 6-25)

The Reassessment describes sources of variability in exposure and intake—for example, contaminated poultry feed, increased consumption of fish, and occupational exposures. This section concludes that a high intake would be about three times the population mean, but again there is no quantification of the MOE or any attempt to link high-exposure groups to specific end points (except for breast-feeding, which is covered in the following section).

Breast-Feeding Infants Have Higher Intakes of TCDD, Other Dioxins, and DLCs for a Short but Developmentally Important Part of Their Lives but the Widely Recognized Benefits of Breast-Feeding Outweigh the Risks

Reassessment (Part III, pp. 6-26 to 6-27)

The Reassessment reiterates the information on breast-feeding and points out that the average daily intake by the infant over the first year of suckling would be 87 times the adult daily intake. It correctly points out that this would not result in an 87-fold higher body burden because of the rapid increase in body weight and more rapid elimination. The Reassessment reiterates the advantages of breast-feeding in general and concludes that reevaluation of TCDD, other dioxins, and DLCs does not alter the previous advice, especially because the risk assessment is based on body burden. While not disagreeing with the conclusion, the committee considers the Reassessment to be superficial on this point. It does not support its position with well-founded evidence, it does not consider the impact of body composition (e.g., percent body fat) on distribution of the body burden in infants, and, most important it makes no attempt to compare the intakes by infants with the doses producing adverse effects in the relevant animal studies (that is, on those involving in utero exposure and subsequent assessment of developmental parameters in early life).

Many Dioxin Sources Have Been Identified and Emissions to the Environment Have Been Reduced

Reassessment (Part III pp. 6-27 to 6-29)

This summary of previously presented information on sources and emissions is adequate. (The committee noted, however, that it is largely irrelevant to this part of the Reassessment because it does not consider or contribute to risk characterization.) See Chapter 4 for the committee's recommendations on sources and emissions.

TCDD, Other Dioxins, and DLCs Dioxins Are Widely Distributed in the Environment at Low Concentrations Primarily as a Result of Air Transport and Deposition

Reassessment (Part III, pp. 6-29 to 6-30)

This summary of previously presented information on sources and emissions is adequate. (The committee noted, however, that it is largely irrelevant to this part of the Reassessment because it does not consider or contribute to risk characterization.)

Environmental Levels, Emissions, and Human Exposures Have Declined During Recent Decades

Reassessment (Part III, p. 6-30)

This summary of previously presented information on sources and emissions is adequate. (The committee noted, however, that it is largely irrelevant to this part of the Reassessment because the data are not interpreted in the context of risk characterization.)

Risk Characterization Summary Statement

Reassessment (Part III, pp. 6-30 to 6-34)

This section provides a reasonable summary of the preceding parts of Chapter 6 of Part III.

CONCLUSIONS AND RECOMMENDATIONS

The committee considered Chapter 6 of Part III of the Reassessment to be the most important section, but in many ways it was the weakest and least scientifically rigorous in its support of the decisions made.

EPA used linear extrapolation from the POD, the ED_{01}, derived from the cancer epidemiological studies to calculate a CSF. The resulting cancer risk estimate of 1×10^{-3} per pg TEQ/kg of body weight per day for both background intakes and incremental intakes above background was considered by EPA to be the most appropriate approach. Using a linear extrapolation approach in the Reassessment was one of the most critical decisions by EPA. Use of this approach was not supported by a scientifically rigorous argument, nor was there a balanced presentation of arguments using the same data to support the calculation and interpretation of an MOE. EPA did not adequately discuss the risk management implications of the cancer risk estimate, which might be interpreted to indicate the need to reduce the current exposure of the general population between 10-fold and 1,000-fold to limit the calculated cancer risk between 1 in 10,000 and 1 in 1,000,000. Such a use and interpretation of the slope factor would require EPA to consider the validity of the linear model over many orders of magnitude.

The Reassessment stated that it used an MOE approach for noncancer effects, but the discussion did not focus on the MOE values for different adverse effects. An important improvement over past EPA practice was the reliance in the Reassessment on an estimated ED (BMD) for noncancer effects rather than on the traditional NOAEL and LOAEL as the POD. An ED can be calculated mathematically from a fitted dose-response model and is not limited to the experimental doses, thus representing a significant advance in dose-response assessment. However, the computation of the ED_{01} for continuous noncancer effects was critical, where the ED_{01} was not the dose associated with a 1% incremental incidence of an adverse effect but was the dose associated with a change in the mean response from the background level that was 1% of the maximum possible total response range. EPA made no attempt to present the biological significance of such changes for each of the different continuous end points of studies subject to dose-response modeling.

The adoption of such a novel approach gave extremely low general MOE values and was used by EPA as justification for not analyzing and interpreting the MOE values for each end point and also for not using the massive TCDD database to identify an RfD.

Because EPA decided not to define an RfD, the Reassessment lacked detailed risk characterization information—for example, the proportion of the population with intakes above the RfD, detailed assessment of population groups, and contributions of the major food sources for those individuals with high intakes. The lack of such a focus in the Reassessment results in a diffuse risk characterization that is difficult to follow and that does not provide clear guidance to risk managers.

The Reassessment should describe clearly the following aspects:

1. The effects seen at the lowest body burdens that are the primary focus for any risk assessment—the "critical effects."

2. The modeling strategy used for each noncancer effect, paying particular attention to the critical effects, and the selection of a point of comparison based on the biological significance of the effect; if the ED_{01} is retained, then the biological significance of the response should be defined and the precision of the estimate given.

3. The precision and uncertainties associated with the body burden estimates for the critical effects at the point of comparison, including the use of total body burden rather than modeling steady-state concentrations for the relevant tissue.

4. The committee encourages EPA to calculate RfDs as part of its effort to develop appropriate margins of exposure for different end points and risk scenarios, including the proportions of the general population and of any identified groups that might be at increased risk (See Table A-1 in the Reassessment, Part III Appendix, for the different effects; appropriate exposure information would need to be generated.) Interpretation of the calculated values should take into consideration the uncertainties in the POD values and intake estimates.

5. Consideration of individuals in susceptible life stages or groups (e.g., children, women of childbearing age, and nursing infants) who might require an estimation of a separate MOE using specific exposure data.

6. Distributions that provide clear insights about the uncertainty in the risk assessments, along with discussion about the key contributors to the uncertainty.

8

Conclusions and Recommendations

CLASSIFICATION OF TCDD AS CARCINOGENIC TO HUMANS

In its charge, the committee was requested to comment specifically on the U.S Environmental Protection Agency (EPA) conclusion that 2,3,7,8-tetrachlorodibenzo-*p*-dioxin (TCDD, also referred to as dioxin) is best characterized as "carcinogenic to humans." Both EPA and the International Agency for Research on Cancer (IARC), an arm of the World Health Organization (WHO), have established criteria for qualitatively classifying chemicals into various categories based on the weight of scientific evidence from animal, human epidemiological, and mechanism or mode-of-action studies. In 1997, an expert panel convened by IARC concluded that the weight of scientific evidence for TCDD carcinogenicity in humans supported its classification as a Class 1 carcinogen—"carcinogenic to humans." In 1985, EPA classified TCDD as a "probable human carcinogen" based on the data available at the time, but in the latest Reassessment (2003),[1] EPA concluded that TCDD was "best characterized as 'carcinogenic to humans." The National Toxicology Program (NTP 2000) also classified TCDD as "known to be a human carcinogen."

After reviewing EPA's 2003 Reassessment and other scientific information and in light of EPA's recently revised 2005 *Guidelines for Carcinogen*

[1] *The Exposure and Human Health Reassessment of 2,3,7,8-Tetrachlorodibenzo-p-dioxin (TCDD) and Related Compounds* (EPA 2003a, Part I; 2003b, Part II; 2003c, Part III) is collectively referred to as the Reassessment.

Risk Assessment (cancer guidelines), the committee concludes that the classification of TCDD as "carcinogenic to humans"—a designation suggesting the greatest degree of certainty about carcinogenicity—versus "likely to be carcinogenic to humans"—the next highest designation—is somewhat subjective and depends largely on the definition and interpretation of the criteria used for classification. The true weight of evidence lies on a continuum, with no obvious point or "bright line" that readily distinguishes between those two categories.

Referring to the specific definitions in EPA's 2005 cancer guidelines for qualitative classification of chemical carcinogens, the NRC committee was split on whether the evidence met *all* the criteria necessary for classification of TCDD as "carcinogenic to humans," although the committee unanimously agreed on a classification of at least "likely to be carcinogenic to humans." The committee concludes that the weight of epidemiological evidence that TCDD is a human carcinogen is not strong, but the human data available from occupational cohorts are consistent with a modest positive association between relatively high body burdens of TCDD and increased mortality from all cancers. Positive animal studies and mechanistic data provide additional support for classification of TCDD as a human carcinogen. The committee recommends that EPA summarize its rationale for concluding that TCDD satisfies the criteria set out in its cancer guidelines for designation as either "carcinogenic to humans" or "likely to be a human carcinogen."

If EPA continues to designate TCDD as "carcinogenic to humans" under the new guidelines, it should explain whether this conclusion reflects a finding that there is a strong association between TCDD exposure and human cancer or between TCDD exposure and key precursor events of TCDD's mode of action (presumably aromatic hydrocarbon receptor [AHR] binding). If its finding reflects the latter association, EPA should explain why that end point (e.g., AHR binding) represents a "key precursor event."

As noted above, the committee concludes that the distinction between these two categories is based more on semantics than on science and recommends that EPA focus its energies and resources on more carefully delineating the assumptions used in quantitative risk estimates for TCDD, other dioxins, and dioxin-like compounds (DLCs) derived from human and animal studies.

The committee agrees that other dioxins and DLCs are most appropriately classified as "likely to be carcinogenic to humans." If EPA continues to classify TCDD as "carcinogenic to humans," more justification will be required to explain why a mixture containing TCDD would not also meet the classification of "carcinogenic to humans."

USE OF LOW-DOSE LINEAR VERSUS THRESHOLD (NONLINEAR) EXTRAPOLATION MODELS FOR QUANTITATIVE CANCER RISK ESTIMATIONS

The committee unanimously agrees that the current weight of evidence for TCDD, other dioxins, and DLCs carcinogenicity favors the use of nonlinear methods for extrapolation below the point of departure (POD) of mathematically modeled human or animal data. However, the committee recognizes that it is not scientifically possible to exclude totally a linear response at doses below the POD, so it recommends that EPA provide risk estimates using both approaches and describe their scientific strengths and weaknesses to inform risk managers of the importance of choosing a linear vs. nonlinear method of extrapolation. To the extent that EPA favors using default assumptions for regulating dioxin as though it were a linear carcinogen, such a conclusion should be made as part of risk management. EPA should strictly adhere to the distinction between risk assessment, which is a scientific activity, and risk management, which takes into account other factors.

USE OF THE 1% RESPONSE LEVEL AS A POINT OF DEPARTURE FOR LOW-DOSE RISK ESTIMATION

The Reassessment adopts the benchmark dose (BMD) method to replace the traditional, less quantitative approach of using no-observed-adverse-effect level (NOAEL) and lowest-observed-adverse-effect level (LOAEL) to characterize noncancer effects. A BMD (or an effective dose) can be calculated mathematically from a fitted dose-response model and is not limited to the experimental doses. The BMD method is a significant advance in dose-response modeling, and EPA's use of BMD is highly commendable. However, the determination of an ED at the 1% response level (ED_{01}) for continuous noncancer effects is not without significant limitation. Specifically, the ED_{01} was the dose associated with a change in mean response away from the background level by 1% of the maximum possible total response range. For some noncancer end points, the significance of such a change may be difficult to identify both clinically and statistically and can be well within the variation of the control data. The biological significance of this magnitude of change represented by the ED_{01} values for different continuous noncancer end points should be evaluated.

The adoption of such a novel approach gave extremely low margin-of-exposure (MOE) values compared with background exposures and was used by EPA as justification for not analyzing and interpreting the MOE values for each end point and also for not using the massive dioxin database to set a reference dose (RfD).

In its evaluation of the ED_{01} used for cancer risk assessment, the committee concluded that EPA had not adequately justified use of the 1% response level as the POD for the analysis of either the epidemiological or the animal bioassay data. Even though it is necessary to demonstrate that the POD is within the range of the observed data, that by itself is not sufficient to justify use of the ED_{01}. Other conditions, such as demonstrating that the POD is relatively insensitive to functional form (as noted in the cancer guidelines [EPA 2005a]), must also be satisfied. EPA should acknowledge the larger extrapolation from justifiable POD values down to environmentally relevant doses that would be necessitated by use of a higher response-level POD.

With regard to EPA's review of the animal cancer bioassay data, the committee recommends that EPA establish clear criteria for the inclusion of different data sets. The reliance on data for one site from one gender of one species, as reported by a single study, does not adequately represent the full range of data available. The committee recommends that EPA consider the full range of data, including the new NTP animal bioassay studies on TCDD, for quantitative dose-response assessment.

For the various noncancer end points, EPA should describe more clearly how and why the ED_{01} values were determined in animals and translated to human equivalents. At the least, the risk assessment should provide more apparent and parallel calculations using a 5% response level as the POD to demonstrate the impact that this assumption might have on both the point estimates of risk at low doses and the range of uncertainty surrounding that point estimate. This recommendation applies to extrapolation for cancer risk estimates, for which an ED_{01} was also used, as well as for noncancer risk estimates.

Although the committee commends EPA's extensive efforts on dose-response modeling of a large number of data sets, particularly those of noncancer end points, it is concerned that selection of the final model for computing POD values was not based on a statistical assessment of model goodness of fit, particularly at low doses. An inadequately fitted model could substantially alter extrapolation to low doses and therefore is a source of error that can result in significant uncertainty. The committee recommends using statistically rigorous methods for assessing model fit to control and reduce this source of uncertainty related to selection of a POD. Although the committee encourages EPA to use thorough statistical analyses of data, it also cautions that "statistical significance" does not always equate with "biological significance," and thus sound scientific judgment, in addition to statistical analysis, is a critical element of data interpretation.

CHARACTERIZATION OF UNCERTAINTY FOR RISK ESTIMATES

Overall, the committee found that the Reassessment qualitatively addressed many sources of uncertainty and variability but that it failed quantitatively to sufficiently address uncertainty and variability that resulted from the numerous decisions EPA made in deriving point estimates of risk in the comprehensive risk assessment. In contrast, EPA used concerns about uncertainties and uncertainty factors as part of the justification for not setting an RfD for noncancer effects (see Chapter 7 for further discussion).

The committee recommends that EPA provide statistical estimates of the upper-, lower-, and central-bound risk estimates for all quantitative risk estimates. In light of the magnitude of this uncertainty, the committee considers identification of a point estimate value for the dioxin cancer slope factor (CSF), even a point estimate designated as an upper bound, to confer a false sense of precision. EPA should identify the sources of uncertainty and quantitatively characterize their impact on the probability distribution that describes the set of plausible CSF values. If necessary, EPA should acknowledge that the information available is not sufficient to support designation of a meaningful point estimate.

The committee recommends that EPA more completely characterize uncertainty associated with cancer risk estimates inferred from the epidemiological data (1) by taking into account the full range of ED values statistically consistent with the data (not only the central and lower estimates); (2) by considering alternative PODs; (3) by considering biologically plausible alternative dose-response functional forms consistent with the data; and (4) by considering uncertainty associated with the half-life estimates of TCDD in humans for the purpose of back-extrapolating exposures in occupational cohort studies.

The Reassessment did not provide details about the magnitudes of the various uncertainties surrounding the decisions that EPA makes about dose metrics (e.g., the impact of species differences in percentage body fat on the steady-state concentrations present in nonadipose tissues). The committee recommends that the Reassessment use simple physiologically based pharmacokinetic models to define and characterize the uncertainty of any differences between humans and rodents in the relationship between total body burden at steady state (as calculated from the intake, half-life, and bioavailability) and tissue concentrations; EPA should modify the estimated human equivalent intakes when necessary. While PBPK modeling may itself introduce uncertainty, the process of building the PBPK model should help to reduce the far greater uncertainty and likelihood of error that arises when PBPK considerations are not included explicitly. Many opportunities exist to further characterize sources of uncertainty and variability related to the dose metric choices, and the committee recommends that EPA improve the Reassessment by providing a clear evaluation of the impacts of possible choices on the risk estimates.

The committee recommends that EPA make greater use of mechanistic information to assess the biological plausibility of different mathematical models, use more rigorous criteria (e.g., goodness-of-fit test) for selecting a model for deriving a POD, and clearly identify the benchmark response level of toxicological significance for noncancer end points.

USE OF TOXIC EQUIVALENCY FACTORS FOR RISK ESTIMATION OF DLCS AND MIXTURES OF DLCS

Overall, even given the inherent uncertainties, the toxic equivalency factor (TEF) method provides a reasonable, scientifically justifiable, and widely accepted method to estimate the relative toxic potency of dioxins, other than TCDD, and DLCs, relative to TCDD, on human and animal health. However, the Reassessment should acknowledge the need for better uncertainty analysis of the TEF values. The committee also supports a previous recommendation from the EPA Science Advisory Board (SAB) "that, as a follow up to the Reassessment, EPA should establish a task force to build consensus probability density functions for the ... chemicals for which TEFs have been established, or to examine related approaches such as those based on fuzzy logic."

USE OF BODY BURDEN AS THE PRIMARY DOSE METRIC FOR CROSS-SPECIES EXTRAPOLATION

Although the committee agrees that use of body burden as the dose metric is the most reasonable and pragmatic approach at the present time, a number of uncertainties in using body burden to develop risk estimates should be addressed. The magnitudes of the various uncertainties are not clearly defined. The most significant impact is the species differences in percentage body fat on the relationship between body burden and the concentrations present in nonadipose tissues. An analysis of the impact of possible uncertainties in the dose metric on the final risk estimates would be informative.

It remains to be determined whether the current WHO TEFs, which were developed to assess the relative toxic potency of a mixture to which an organism is directly exposed by dietary intake, are appropriate for body burden toxic equivalent quotient (TEQ) determinations, which are derived from the concentrations of different congeners measured in body fat. If body burdens are to be used as the dose metric, a separate set of body burden TEFs should be developed and applied for this evaluation. Without these corrected values, the overall TEQs estimated by use of intake TEFs might be substantially in error.

EPA'S EXPOSURE ASSESSMENT OF TCDD, OTHER DIOXINS, AND DLCS IN THE UNITED STATES

To assess total emissions of TCDD, other dioxins, and DLCs, EPA used a "bottom-up" approach, which attempted to identify all source categories and then estimated the magnitude of emissions for each category. However, a "top-down" approach that attempts to account for the levels measured in receptors (e.g., people, animals, and plants) could give rise to substantially different information. Such alternative approaches are likely to give rise to significantly different estimates of the historical levels of dioxin emissions. Both approaches come with uncertainties, and EPA could benefit significantly from using them simultaneously to set plausible bounds on the historical and current trends in emissions.

Although beyond the scope of the review of the EPA Reassessment, the committee noted that it would be useful for EPA to set up an active congener-specific database of typical concentrations for the whole range of polychlorinated dibenzo-*p*-dioxins (PCDDs), polychlorinated dibenzofurans (PCDFs), and dioxin-like polychlorinated biphenyls (PCBs) (included in the WHO TEF list) present in food. This database should be based on a compendium of all available data that would be updated on a regular basis with new data as they are published in the peer-reviewed literature. Such a database should have clear requirements of data quality and traceability (e.g., chemical analysis, representative and targeted sampling, representative of consumer exposure, presentation of data, and handling and presentation of nondetects).

The committee suggests that in the future EPA define a strategy for collecting samples and reanalyzing archived samples to answer a number of remaining questions about exposure trends and to fill some important data gaps.

EPA'S EVALUATION OF IMMUNOTOXICITY OF TCDD, OTHER DIOXINS, AND DLCS

Present clinical findings are inconclusive about whether or in what way TCDD, other dioxins, and DLCs are immunotoxic in humans, and EPA acknowledges that human data are sparse. A series of studies from a Dutch children's cohort showed an association between prenatal exposure to DLCs and changes in immune status. The effects were modest, and laboratory values did not fall significantly outside the full range of normal. Some clinically relevant adverse effects seen in this perinatal study are also seen at higher levels of exposure, although these do not seem to persist. A number of animal studies suggest that the developing immune system is especially sensitive. In light of the large database showing that TCDD, other dioxins, and DLCs are immunotoxic in laboratory animal studies—together with limited human data—EPA is prudent in concluding that these compounds are likely to be human immunotoxicants in the absence of more definitive human data.

However, EPA's conclusion that TCDD and related compounds are immunotoxic at "some dose level" by itself is inadequate. At a minimum, a section or paragraph should be added that discusses the immunotoxicology of TCDD, other dioxins, and DLCs in the context of current AHR biology.

Likewise, some discussion should also be included on the strengths and weaknesses of using genetically homogeneous inbred mice to characterize immunotoxicological risk in the genetically variable human population. Expanding the discussion to include the above crucial points would provide additional balance to Part III, Integrated Summary and Risk Characterization.

Additional comments and recommendations relating to the use of specific data sets for risk assessment of the immunotoxic effects of TCDD, other dioxins, and DLCs are provided in Chapter 6.

EPA'S EVALUATION OF REPRODUCTIVE AND DEVELOPMENTAL TOXICITY OF TCDD, OTHER DIOXINS, AND DLCS

As clearly described in the Reassessment, embryonic and fetal development and reproductive effects are sensitive end points of TCDD toxicity in rodents. It is clear that the fetal rodent is more sensitive than the adult rodent to adverse effects of TCDD. Comparable human data are generally lacking, and the sensitivity of humans to these end points is less apparent.

The committee recommends that EPA address more thoroughly how the effective doses used in the animal pregnancy models relate to human reproductive and developmental toxicity and risk information, including TEFs and TEQs. For available human clinical data on reproductive and developmental end points, EPA should establish formal principles of, and a formal mechanism for, evidence-based classification and systematic statistical review, including meta-analysis when possible.

Finally, EPA should discuss the dose-response effects of TCDD, other dioxins, and DLCs on the adult female reproductive system that result in endocrine disruption in animals. Based on the dose-response data provided in these studies, the impact on human risk assessment should be presented.

EPA'S EVALUATION OF OTHER TOXIC END POINTS

In general, the committee determined that the Reassessment adequately addressed the available data on whether exposures to TCDD, other dioxins, and DLCs are likely to be significant risk factors for other toxic end points, such as chloracne, thyroid function, liver function, diabetes, lipid disorders, and cardiovascular diseases. In humans, the relationship between dioxin exposure and risk of individual, clinically significant, noncancer end points remains uncertain, except for chloracne.

The overall conclusions in the Reassessment about noncancer risks due to exposure to TCDD, other dioxins, and DLCs are, in general, cautiously stated, and the uncertainty of suspected relationships is acknowledged. Nonetheless, the limitations of individual human studies are not uniformly addressed, and the broad 95% confidence intervals accompanying some reported statistically significant effects are not discussed in the context of the uncertainty that these broad confidence limits imply. Conversely, statistically nonsignificant effects are sometimes highlighted, presenting an implied potential for unobserved detrimental effects without a firm evidence base. For available human clinical data for other noncancer end points, EPA should establish formal principles of, and a formal mechanism for, evidence-based classification and systematic statistical review, including meta-analysis when possible.

With respect to human noncancer end points, the committee determined that the Reassessment text should be revised to include the relevant, more recent data and, when appropriate, the quality and data uncertainty of the studies referenced. When the mechanism is established, currently available and newly available human clinical studies should be subject to such systematic review and formal evidence-based assessment. The quality of the available evidence should be reported, and the strength or weakness of a presumptive association should be classified according to currently accepted criteria for levels of evidence.

New studies on effects of TCDD on the developing vascular system suggest that this system could be a highly sensitive target and suggest that this area be identified as an important data gap in the understanding of the potential adverse effects of TCDD, other dioxins, and DLCs.

EPA'S OVERALL APPROACH TO RISK CHARACTERIZATION

As discussed above, EPA used linear extrapolation from the POD (the ED_{01}) derived from the cancer epidemiological studies and animal bioassays to calculate a CSF. The selection of the default linear extrapolation approach was one of the most critical decisions in the Reassessment, but the decision to use this approach was not supported by a scientifically rigorous argument, nor was there a balanced presentation of arguments that would support the calculation and interpretation of a MOE with the same data. The committee determined that a balanced presentation of available data could support the use of a nonlinear model consistent with a receptor-mediated mode of action with subsequent calculations and interpretation of

MOE values. (For cancer risk assessment, the threshold approach should be used in addition to the linear approach.)

Because EPA decided not to define an RfD, the Reassessment lacked detailed risk characterization information—for example, the proportion of the population with intakes above the RfD; detailed assessment of population groups, such as those with occupational exposures; and the contributions of major food sources and other environmental sources for those people with high intakes. The lack of such a focus in the Reassessment results in a diffuse risk characterization that is difficult to follow and that does not provide clear advice to risk managers.

The committee points out particular areas that could be improved in Part III of the Reassessment. In particular, the risk characterization chapter of the Reassessment should describe concisely and clearly the following aspects.

1. The effects seen at the lowest body burdens that are the primary focus for any risk assessment—the "critical effects."
2. The modeling strategy used for each noncancer effect modeled, paying particular attention to the critical effects, and the selection of a point of comparison based on the biological significance of the effect; if the ED_{01} is retained, then the biological significance of the response should be defined and the precision of the estimate given.
3. The precision and uncertainties associated with the body burden estimates for the critical effects at the point of comparison, including the use of total body burden rather than modeling steady-state concentrations for the relevant tissue.
4. The committee encourages EPA to calculate RfDs as part of its effort to develop appropriate margins of exposure for different end points and risk scenarios, including the proportions of the general population and of any identified groups that might be at increased risk (See Table A-1 in the Reassessment, Part III Appendix, for the different effects; appropriate exposure information would need to be generated.) Interpretation of the calculated values should take into consideration the uncertainties in the POD values and intake estimates.
5. Consideration of individuals in susceptible life stages or groups (e.g., children, women of childbearing age, and nursing infants) who might require estimation of a separate MOE using specific exposure data.
6. Distributions that provide clear insights about the uncertainty in the risk assessments, along with discussion of the key contributors to the uncertainty.

The committee recognizes that it will require a substantial amount of effort by EPA to incorporate all the changes recommended in this review;

however, it does not advocate a substantial expansion in the length of the Reassessment. Rather, the committee encourages EPA to address the major concerns raised in this review and to finalize the current Reassessment as quickly, efficiently, and concisely as possible. The committee agreed that it is important for EPA to recognize that new advances in the understanding of the toxicity of TCDD, other dioxins, and DLCs could require reevaluation of key assumptions in the risk assessment document. The committee recommends that EPA routinely monitor new scientific information with the understanding that future revisions may be required to maintain a risk assessment that is based on current state-of-the-science. However, the committee also recognizes that stability in regulatory policy is important to the regulated community and thus expects that science-based changes in regulatory policy on TCDD, other dioxins, and DLCs will be invoked only in the face of compelling new information that would warrant revision of its final risk assessment. Such substantial gains in knowledge are not likely to occur frequently.

References

Abraham, K., O. Papke, A. Gross, O. Kordonouri, S. Wiegand, U. Wahn, and H. Helge. 1998. Time course of PCDD/PCDF/PCB concentrations in breast-feeding mothers and their infants. Chemosphere 37(9-12):1731-1741.

Alaluusua, S., P.L. Lukinmaa, T. Vartiainen, M. Partanen, J. Torppa, and J. Tuomisto. 1996. Polychlorinated dibenzo-*p*-dioxins and dibenzofurans via mother's milk may cause developmental defects in the child's teeth. Environ. Toxicol. Pharmacol. 1(3):193-197.

Alaluusua, S., P.L. Lukinmaa, J. Torppa, J. Tuomisto, and T. Vartiainen. 1999. Developing teeth as biomarker of dioxin exposure. Lancet 353(9148):206.

Allen, J.R., D.A. Barsotti, J.P. VanMiller, L.J. Abrahamson, and J.J. Lalich. 1977. Morphological changes in monkeys consuming a diet containing low levels of 2,3,7,8-tetrachlorodibenzo-*p*-dioxin. Food Cosmet. Toxicol. 15(5):401-410.

Amakura, Y., T. Tsutsumi, M. Nakamura, H. Kitagawa, J. Fujino, K. Sasaki, T. Yoshida, and M. Toyoda. 2002. Preliminary screening of the inhibitory effect of food extracts on activation of the aryl hydrocarbon receptor induced by 2,3,7,8-tetrachlorodibenzo-*p*-dioxin. Biol. Pharm. Bull. 25(2):272-274.

Andersson, P., J. McGuire, C. Rubio, K. Gradin, M.L. Whitelaw, S. Pettersson, A. Hanberg, and L. Poellinger. 2002. A constitutively active dioxin/aryl hydrocarbon receptor induces stomach tumors. Proc. Natl. Acad. Sci. USA 99(15):9990-9995.

Antkiewicz, D.S., C.G. Burns, S.A. Carney, R.E. Peterson, and W. Heideman. 2005. Heart malformation is an early response to TCDD in embryonic zebrafish. Toxicol. Sci. 84(2): 368-377.

Anttila, S., X.D. Lei, E. Elovaara, A. Karjalainen, W. Sun, H. Vainio, and O. Hankinson. 2000. An uncommon phenotype of poor inducibility of CYP1A1 in human lung is not ascribable to polymorphisms in the AHR, ARNT, or CYP1A1 genes. Pharmacogenetics 10(8):741-751.

Astroff, B., T. Zacharewski, S. Safe, M.P. Arlotto, A. Parkinson, P. Thomas, and W. Levin. 1988. 6-Methyl-1,3,8-trichlorodibenzofuran as a 2,3,7,8-tetrachlorodibenzo-*p*-dioxin antagonist: Inhibition of the induction of rat cytochrome P-450 isozymes and related monooxygenase activities. Mol. Pharmacol. 33(2):231-236.

ATSDR (Agency for Toxic Substances and Disease Registry). 1998. Toxicological Profile for Chlorinated Dibenzo-p-Dioxins. Agency for Toxic Substances and Disease Registry, Public Health Service, U.S. Department of Health and Human Services, Atlanta, GA. December 1998 [online]. Available: http://www.atsdr.cdc.gov/toxprofiles/tp104-p.pdf [accessed July 5, 2005].

ATSDR (Agency for Toxic Substances and Disease Registry). 2000. Toxicological Profile for Polychlorinated Biphenyls (PCBs). Agency for Toxic Substances and Disease Registry, Public Health Service, U.S. Department of Health and Human Services, Atlanta, GA. November 2000 [online]. Available: http://www.atsdr.cdc.gov/toxprofiles/tp17.html [accessed Sept. 2, 2005].

Axmon, A., L. Rylander, U. Stromberg, and L. Hagmar. 2004. Altered menstrual cycles in women with a high dietary intake of persistent organochlorine compounds. Chemosphere 56(8):813-819.

Aylward, L.L., R.C. Brunet, G. Carrier, S.M. Hays, C.A. Cushing, L.L. Needham, D.G. Patterson, Jr., P.M. Gerthoux, P. Brambilla, and P. Mocarelli. 2005. Concentration-dependent TCDD elimination kinetics in humans: Toxicokinetic modeling for moderately to highly exposed adults from Seveso, Italy, and Vienna, Austria, and impact on dose estimates for the NIOSH cohort. J. Expo. Anal. Environ. Epidemiol. 15(1):51-65.

Baars, A.J., M.I. Bakker, R.A. Baumann, P.E. Boon, J.L. Freijer, L.A. Hoogenboom, R. Hoogerbrugge, J.D. van Klaveren, A.K. Liem, W.A. Traag, and J. de Vries. 2004. Dioxins, dioxin-like PCBs and non-dioxin-like PCBs in foodstuffs: Occurrence and dietary intake in The Netherlands. Toxicol. Lett. 151(1):51-61.

Bailey, G.S., J.D. Hendricks, D.W. Shelton, J.E. Nixon, and N.E. Pawlowski. 1987. Enhancement of carcinogenesis by the natural anticarcinogen indole-3-carbinol. J. Natl. Cancer Inst. 78(5):931-934.

Bannister, R., D. Davis, T. Zacharewski, I. Tizard, and S. Safe. 1987. Aroclor 1254 as a 2,3,7,8-tetrachlorodibenzo-*p*-dioxin antagonist: Effects on enzyme induction and immunotoxicity. Toxicology 46(1):29-42.

Bannister, R., L. Biegel, D. Davis, B. Astroff, and S. Safe. 1989. 6-Methyl-1,3,8-trichlorodibenzofuran (MCDF) as a 2,3,7,8-tetrachlorodibenzo-p-dioxin antagonist in C57BL/6 mice. Toxicology 54(2):139-150.

Barsotti, D.A., L.J. Abrahamson, and J.R. Allen. 1979. Hormonal alterations in female rhesus monkeys fed a diet containing 2,3,7,8-tetrachlorodibenzo-*p*-dioxin. Bull. Environ. Contam. Toxicol. 21(4-5):463-469.

Becher, H., K. Steindorf, and D. Flesch-Janys. 1998. Quantitative cancer risk assessment for dioxins using an occupational cohort. Environ. Health Perspect. 106(Suppl. 2):663-670.

Bertazzi, P.A., I. Bernucci, G. Brambilla, D. Consonni, and A.C. Pesatori. 1998. The Seveso studies on early and long-term effects of dioxin exposure: A review. Environ. Health Perspect. 106(Suppl. 2):625-633.

Bhattacharyya, K.K., P.B. Brake, S.E. Eltom, S.A. Otto, and C.R. Jefcoate. 1995. Identification of a rat adrenal cytochrome P450 active in polycyclic hydrocarbon metabolism as rat CYP1B1. Demonstration of a unique tissue-specific pattern of hormonal and aryl hydrocarbon receptor-linked regulation. J. Biol. Chem. 270(19):11595-11602.

Biegel, L., M. Harris, D. Davis, R. Rosengren, L. Safe, and S. Safe. 1989. 2,2'4,4'5,5'-hexachlorobiphenyl as a 2,3,7,8-tetrachlorodibenzo-*p*-dioxin antagonist in C57BL/6J mice. Toxicol. Appl. Pharmacol. 97(3):561-571.

Birmingham, B. 1990. Analysis of PCDD and PCDF patterns in soil samples: Use in the estimation of the risk of exposure. Chemosphere 20(7-9):807-814.

Birnbaum, L.S., and L.A. Couture. 1988. Disposition of octachlorodibenzo-*p*-dioxin (OCDD) in male rats. Toxicol. Appl. Pharmacol. 93(1):22-30.

Birnbaum, L.S., R.E. Morrissey, and M.W. Harris. 1991. Teratogenic effects of 2,3,7,8-tetrabromodibenzo--dioxin and three polybrominated dibenzofurans in C57BL/6N mice. Toxicol. Appl. Pharmacol. 107(1):141-152.

Birnbaum, L.S., D.F. Staskal, and J.J. Diliberto. 2003. Health effects of polybrominated dibenzo-*p*-dioxins (PBDDs) and dibenzofurans (PBDFs). Environ. Int. 29(6):855-860.

Blanck, H.M., M. Marcus, C. Rubin, P.E. Tolbert, V.S. Hertzberg, A.K. Henderson, and R.H. Zhang. 2002. Growth in girls exposed in utero and postnatally to polybrominated biphenyls and polychlorinated biphenyls. Epidemiology 13(2):205-210.

Bradfield, C.A., and A. Poland. 1988. A competitive binding assay for 2,3,7,8-tetrachlorodibenzo-*p*-dioxin and related ligands of the Ah receptor. Mol. Pharmacol. 34(5): 682-688.

Breivik, K., A. Sweetman, J.M. Pacyna, and K.C. Jones. 2002a. Towards a global historical emission inventory for selected PCB congeners - a mass balance approach. 1. Global production and consumption. Sci. Total Environ. 290(1-3):181-198.

Breivik, K., A. Sweetman, J.M. Pacyna, and K.C. Jones. 2002b. Towards a global historical emission inventory for selected PCB congeners - a mass balance approach. 2. Emissions. Sci. Total Environ. 290(1-3):199-224.

Brown, M.M., U.A. Schneider, J.R. Petrulis, and N.J. Bunce. 1994. Additive binding of polychlorinated biphenyls and 2,3,7,8-tetrachlorodibenzo-*p*-dioxin to the murine hepatic Ah receptor. Toxicol. Appl. Pharmacol. 129(2):243-251.

Brzuzy, L.P., and R.A. Hites. 1995. Estimating the atmospheric deposition of polychlorinated dibenzo-*p*-dioxins and dibenzofurans from soils. Environ. Sci. Technol. 29(8):2090-2098.

Brzuzy, L.P., and R.A. Hites. 1996. Global mass balance for polychlorinated dibenzo-*p*-dioxins and dibenzofurans. Environ. Sci. Technol. 30(6):1797-1804.

Bueno de Mesquita, H.B., G. Doornbos, D.A. Van der Kuip, M. Kogevinas, and R. Winkelmann. 1993. Occupational exposure to phenoxy herbicides and chlorophenols and cancer mortality in the Netherlands. Am. J. Ind. Med. 23(2):289-300.

Burleson, G.R., H. Lebrec, Y.G. Yang, J.D. Ibanes, K.N. Pennington, and L.S. Birnbaum. 1996. Effect of 2,3,7,8-tetrachlorodibenzo-*p*-dioxin (TCDD) on influenza virus host resistance in mice. Fundam. Appl. Toxicol. 29(1):40-47.

Busser, M.T., and W.K. Lutz. 1987. Stimulation of DNA synthesis in rat and mouse liver by various tumor promoters. Carcinogenesis 8(10):1433-1437.

Carney, S.A., R.E. Peterson, and W. Heideman. 2004. 2,3,7,8-Tetrachlorodibenzo-*p*-dioxin activation of the aryl hydrocarbon receptor/aryl hydrocarbon receptor nuclear translocator pathway causes developmental toxicity through a CYP1A-independent mechanism in zebrafish. Mol. Pharmacol. 66(3):512-521.

Cauchi, S., I. Stucker, S. Cenee, P. Kremers, P. Beaune, and L. Massaad-Massade. 2003. Structure and polymorphisms of human aryl hydrocarbon receptor repressor (AhRR) gene in a French population: Relationship with CYP1A1 inducibility and lung cancer. Pharmacogenetics 13(6):339-347.

CDC (Centers for Disease Control). 1988. Health status of Vietnam veterans. II. Physical health. JAMA 259(18):2708-2714.

CDEP (Connecticut Department of Environmental Protection). 1988. Measurement of Selected Polychlorinated Dibenzo-*p*-Dioxins and Polychlorinated Dibenzofurans in Ambient Air in the Vicinity of Wallingford, Connecticut. Project No. 7265-001-004. Prepared for Air Compliance Unit, Connecticut Department of Environmental Protection, Hartford, CT, by ERT, Concord, MA. July 8, 1988.

CDEP (Connecticut Department of Environmental Protection). 1995. Ambient Monitoring for PCDDs/PCDFs in Connecticut-Fall 1993 through Summer 1994, Final report. ENSR Document No. 6350-009-500-R1. ENSR, Acton, MA. September 1995.

CEBM (Centre for Evidence Based Medicine). 2005. Levels of Evidence and Grades of Recommendation. Oxford-Centre for Evidence Based Medicine, Institute of Health Sciences, Headington, Oxford, UK [online]. Available: http://www.cebm.net/levels_of_evidence.asp [accessed July 8, 2005].

CFSAN (Center for Food Safety and Applied Nutrition). 2005a. Dioxin Analysis Results/Exposure Estimates. Center for Food Safety and Applied Nutrition-Office of Plant and Dairy Foods, U.S. Food and Drug Administration [online]. Available: http://www.cfsan.fda.gov/~lrd/dioxdata.html [accessed July 15, 2005].

CFSAN (Center for Food Safety and Applied Nutrition). 2005b. PCDD/PCDF Exposure Estimates. Center for Food Safety and Applied Nutrition-Office of Plant and Dairy Foods, U.S. Food and Drug Administration [online]. Available: http://www.cfsan.fda.gov/~lrd/dioxee.html [accessed July 15, 2005].

Chaffin, C.L., R.L. Stouffer, and D.M. Duffy. 1999. Gonadotropin and steroid regulation of steroid receptor and aryl hydrocarbon receptor messenger ribonucleic acid in macaque granulasa cells during the periovulatory interval. Endocrinology 140(10):4753-4760.

Charnley, G., and J. Doull. 2005. Human exposure to dioxins from food, 1999-2002. Food Chem. Toxicol. 43(5):671-679.

Chen, G., and N.J. Bunce. 2004. Interaction between halogenated aromatic compounds in the Ah receptor signal transduction pathway. Environ. Toxicol. 19(5):480-489.

Chen, I., N. Harper, and S. Safe. 1995. Inhibition of TCDD-induced responses in B6C3F1 mice and hepat1c1c7 cells by indole-3-carbinol. Organohalogen Compounds 25:57-60.

Chen, I., S. Safe, and L. Bjeldanes. 1996. Indole-3-carbinol and diindolylmethane as aryl hydrocarbon (Ah) receptor agonists and antagonists in T47D human breast cancer cells. Biochem. Pharmacol. 51(8):1069-1076.

Cleverly, D., M. Monetti, L. Phillips, P. Cramer, M. Heit, S. McCarthy, K. O'Rourke, J. Stanley, and D. Winters. 1996. A time-trends study of the occurrences and levels of CDDs, CDFs, and dioxin-like PCBs in sediment cores from 11 geographically distributed lakes in the United States. Organohalogen Compounds 28:77-82.

Cleverly, D.H., D. Winters, J. Ferrario, J. Schaum, G. Schweer, J. Buchert, C. Greene, A. Dupuy, and C. Byrne. 2000. The national dioxin air monitoring network (NDAMN): Results of the first year of atmospheric measurements of CDDs, CDFs and dioxin-like PCBs in rural and agricultural areas of the United States: June 1998-June 1999. Organohalogen Compounds 45:248-251 [online]. Available: http://www.epa.gov/ncea/pdfs/dioxin/dei/NDAMN_PAPER3a.pdf [accessed August 23, 2005].

Conolly, R.B., and M.E. Andersen. 1997. Hepatic foci in rats after diethylnitrosamine initiation and 2,3,7,8-tetrachlorodibenzo-*p*-dioxin promotion: Evaluation of a quantitative two-cell model and of CYP 1A1/1A2 as a dosimeter. Toxicol. Appl. Pharmacol. 146(2):-281-293.

Cousins, I.T., and D. Mackay. 2001. Strategies for including vegetation compartments in multimedia models. Chemosphere 44(4):643-654.

Cousins, I.T., D. Mackay, and K.C. Jones. 1999a. Measuring and modeling the vertical distribution of semi-volatile organic compounds in soils. II. Model development. Chemosphere 39(14):2519-2534.

Cousins, I.T., B. Gevao, and K.C. Jones. 1999b. Measuring and modeling the vertical distribution of semi-volatile organic compounds in soils. I. PCB and PAH soil core data. Chemosphere 39(14):2507-2518.

Couture, L.A., M.R. Elwell, and L.S. Birnbaum. 1988. Dioxin-like effects observed in male rats following exposure to octachlorodibenzo-*p*-dioxin (OCDD) during a 13-week study. Toxicol. Appl. Pharmacol. 93(1):31-46.

Crump, K.S. 1984. A new method for determining allowable daily intakes. Fundam. Appl. Toxicol. 4(5):854-871.

Crump, K.S., R. Canady, and M. Kogevinas. 2003. Meta-analysis of dioxin cancer dose response for three occupational cohorts. Environ. Health Perspect. 111(5):681-687.

Cummings, A.M., J.L. Metcalf, and L. Birnbaum. 1996. Promotion of endometriosis by 2,3,7,8-tetrachlorodibenzo-*p*-dioxin in rats and mice: Time-dose dependence and species comparison. Toxicol. Appl. Pharmacol. 138(1):131-139.

Dashwood, R.H. 1998. Indole-3-carbinol: Anticarcinogen or tumor promoter in brassica vegetables? Chem. Biol. Interact. 110(1-2):1-5.

Dashwood, R.H., A.T. Fong, D.E. Williams, J.D. Hendricks, and G.S. Bailey. 1991. Promotion of aflatoxin B1 carcinogenesis by the natural tumor modulator indole-3-carbinol: Influence of dose, duration, and intermittent exposure on indole-3-carbinol promotional potency. Cancer Res. 51(9):2362-2365.

Dasmahapatra, A.K., B.A. Wimpee, A.L. Trewin, C.F. Wimpee, J.K. Ghorai, and R.J. Hutz. 2000. Demonstration of 2,3,7,8-tetrachlorodibenzo-*p*-dioxin attenuation of P450 steroidogenic enzyme mRNAs in rat granulosa cell in vitro by competitive reverse transcriptase-polymerase chain reaction. Mol. Cell. Endocrinol. 164(1-2):5-18.

Davis, D., and S. Safe. 1988. Immunosuppressive activities of polychlorinated dibenzofuran congeners: Quantitative structure-activity relationships and interactive effects. Toxicol. Appl. Pharmacol. 94(1):141-149.

DeCaprio, A.P., D.N. McMartin, P.W. O'Keefe, R. Rej, J.B. Silkworth, and L.S. Kaminsky. 1986. Subchronic oral toxicity of 2,3,7,8-tetrachlorodibenzo-p-dioxin in the guinea pig: Comparisons with a PCB-containing transformer fluid pyrolysate. Fundam. Appl. Toxicol. 6(3):454-463.

Della Porta, G., T.A. Dragani, and G. Sozzi. 1987. Carcinogenic effects of infantile and long term 2,3,7,8-tetrachlorodibenzo-*p*-dioxin treatment in the mouse. Tumori 73(2):99-107.

Denison, M.S., and S.R. Nagy. 2003. Activation of the aryl hydrocarbon receptor by structurally diverse exogenous and endogenous chemicals. Ann. Rev. Pharmacol. Toxicol. 43: 309-334.

Denison, M.S., A. Pandini, S.R. Nagy, E.P. Baldwin, and L. Bonati. 2002. Ligand binding and activation of the Ah receptor. Chem. Biol. Interact. 141(1-2):3-24.

DeVito, M.J., T. Thomas, E. Martin, T.H. Umbreit, and M.A. Gallo. 1992. Antiestrogenic action of 2,3,7,8-tetrachlorodibenzo-*p*-dioxin: Tissue-specific regulation of estrogen receptor in CD1 mice. Toxicol. Appl. Pharmacol. 113(2): 284-292.

DeVito, M.J., J.J. Diliberto, D.G. Ross, M.G. Menache, and L.S. Birnbaum. 1997. Dose-response relationships for polyhalogenated dioxins and dibenzofurans following subchronic treatment in mice. I. CYP1A1 and CYP1A2 enzyme activity in liver, lung, and skin. Toxicol. Appl. Pharmacol. 147(2):267-280.

DeVito, M.J., D.G. Ross, A.E. Dupuy, Jr., J. Ferrario, D. McDaniel, and L.S. Birnbaum. 1998. Dose-response relationships for disposition and hepatic sequestration of polyhalogenated dibenzo-*p*-dioxins, dibenzofurans, and biphenyls following subchronic treatment in mice. Toxicol. Sci. 46(2):223-234.

DeVito, M.J., L.S. Birnbaum, and N.J. Walker. 2003. The influence of chemical impurity on estimating relative potency factors for PCBs. Organohalogen Compounds 65:288-291.

DiGiovanni, J., A. Viaje, D.L. Berry, T.J. Slaga, and M.R. Juchau. 1977. Tumor-initiating ability of 2,3,7,8-tetrachlorodibenzo-*p*-dioxin (TCDD) and Arochlor 1254 in the two-stage system of mouse skin carcinogenesis. Bull. Environ. Contam. Toxicol. 18(5): 552-557.

DOD (U.S. Department of Defense). 2005. Air Force Health Study, Final Report. An Epidemiologic Investigation of Health Effects in Air Force Personnel Following Exposure to Herbicides 2002 Follow-up Examination Results May 2002 to March 2005. July 8, 2005 [online]. Available: http://www.brooks.af.mil/AFRL/HED/hedb/default.html [accessed July 29, 2005].

Dong, W., H. Teraoka, K. Yamazaki, S. Tsukiyama, S. Imani, T. Imagawa, J.J. Stegeman, R.E. Peterson, and T. Hiraga. 2002. 2,3,7,8-tetrachlorodibenzo-*p*-dioxin toxicity in the zebrafish embryo: Local circulation failure in the dorsal midbrain is associated with increased apoptosis. Toxicol. Sci. 69(1):191-201.

Dong, W., H. Teraoka, Y. Tsujimoto, J.J. Stegeman, and T. Hiraga. 2004. Role of aryl hydrocarbon receptor in mesencephalic circulation failure and apoptosis in zebrafish embryos exposed to 2,3,7,8-tetrachlorodibenzo-*p*-dioxin. Toxicol. Sci. 77(1):109-116.

Eastin, W.C., J.K. Haseman, J.F. Mahler, and J.R. Bucher. 1998. The National Toxicology Program evaluation of genetically altered mice as predictive models for identifying carcinogens. Toxicol. Pathol. 26(4):461-473.

Eaton, D.L., and C.D. Klaassen. 2001. Principles of toxicology. Pp. 11-34 in Casarett and Doull's Toxicology: The Basic Science of Poisons, 6th Ed., C.D. Klaassen, ed. New York: McGraw-Hill.

Egeland, G.M., M.H. Sweeney, M.A. Fingerhut, K.K. Wille, T.M. Schnorr, and W.E. Halperin. 1994. Total serum testosterone and gonadotropins in workers exposed to dioxin. Am. J. Epidemiol. 139(3):272-281.

Eisenberg, J.N.S., D.H. Bennett, and T.E. McKone. 1998. Chemical dynamics of persistent organic pollutants: A sensitivity analysis relating soil concentration levels to atmospheric emissions. Environ. Sci. Technol. 32(1):115-123.

Ema, M., N. Ohe, M. Suzuki, J. Mimura, K. Sogawa, S. Ikawa, and Y. Fujii-Kuriyama. 1994. Dioxin binding activities of polymorphic forms of mouse and human aryl hydrocarbon receptors. J. Biol. Chem. 269(44):27337-27343.

Emond, C., L.S. Birnbaum, and M.J. DeVito. 2004. Physiologically based pharmacokinetic model for developmental exposures to TCDD in the rat. Toxicol. Sci. 80(1):115-133.

Enan, E., B. Lasley, D. Stewart, J. Overstreet, and C.A. VandeVoort. 1996a. 2,3,7,8-tetrachlorodibenzo-*p*-dioxin (TCDD) modulates function of human luteinizing granulosa cells via cAMP signaling and early reduction of glucose transporting activity. Reprod. Toxicol. 10(3):191-198.

Enan, E., F. Moran, C.A. VandeVoort, D.R. Stewart, J.W. Overstreet, and B.L. Lasley. 1996b. Mechanism of toxic action of 2,3,7,8-tetrachlorodibenzo-*p*-dioxin (TCDD) in cultured human luteinized granulosa cells. Reprod. Toxicol. 10(6): 497-508.

Enzmann, H., E. Bomhard, M. Iatropoulos, H.J. Ahr, G. Schlueter, and G.M. Williams. 1998. Short- and intermediate-term carcinogenicity testing: A review. Part 1. The prototypes mouse skin tumour assay and rat liver focus assay. Food Chem. Toxicol. 36(11): 979-995.

EPA (U.S. Environmental Protection Agency). 1985. Soil Screening Survey at Four Midwestern Sites. EPA 905/4-85-005. Environmental Services Division, Eastern District Office, Region V, U.S. Environmental Protection Agency, Westlake, OH.

EPA (U.S. Environmental Protection Agency). 1987. Interim Procedures for Estimating Risks Associated with Exposures to Mixtures of Chlorinated Dibenzo-p-dioxins and -dibenzofurans (CDDs and CDFs). EPA/625/3-87/012. Risk Assessment Forum, U.S. Environmental Protection Agency, Washington, DC.

EPA (U.S. Environmental Protection Agency). 1992. National Study of Chemical Residues in Fish. EPA 823-R-92-008. Standards and Applied Science Division, Office of Science and Technology, U.S. Environmental Protection Agency, Washington, DC. September 1992.

EPA (U.S. Environmental Protection Agency). 1994. Health Assessment Document for 2,3,7,8-Tetrachlorodibenzo-*p*-dioxin (TCDD) and Related Compounds, Vols. I-III. External Review Draft. EPA/600/BP-92/001a,b,c Office of Health and Environmental Assessment, Office of Research and Development, U.S. Environmental Protection Agency, Washington, DC. August 1994 [online]. Available: http://www.cqs.com/epa/health/ [accessed April 14, 2006].

EPA (U.S. Environmental Protection Agency). 1996a. Columbus Waste-to-Energy Municipal Incinerator Dioxin Soil Sampling Project. Region V, U.S. Environmental Protection Agency, Chicago, IL. April 1996.

EPA (U.S. Environmental Protection Agency). 1998. Assessment of Thyroid Follicular Cell Tumors. EPA/630/R-97/002. Risk Assessment Forum, U.S. Environmental Protection Agency, Washington, DC [online]. Available: http://cfpub.epa.gov/ncea/cfm/recordisplay.cfm?PrintVersion=True&deid= 13102 [accessed July 7, 2005].

EPA (U.S. Environmental Protection Agency). 2000a. Characterization of Dioxins, Furans and PCBs in Soil Samples Collected from the Denver Front Range Area. Region VIII, U.S. Environmental Protection Agency, Denver, CO. October 2000.

EPA (U.S. Environmental Protection Agency). 2000b. Benchmark Dose Technical Guidance Document. External Review Draft. EPA/630/R-00/001. Risk Assessment Forum, U.S. Environmental Protection Agency, Washington, DC. October 2000 [online]. Available: http://www.epa.gov/nceawww1/pdfs/bmds/BMD-External_10_13_2000.pdf [accessed Sept. 1, 2005].

EPA (U.S. Environmental Protection Agency). 2002. Workshop on the Benefits of Reductions in Exposure to Hazardous Air Pollutants: Developing Best Estimates of Dose-Response Functions. EPA-SAB-EC-WKSHP-02-001. EPA Science Advisory Board, Washington, DC. January 2002 [online]. Available: http://www.epa.gov/science1/fiscal02.htm. [accessed July 6, 2005].

EPA (U.S. Environmental Protection Agency). 2003a. Exposure and Human Health Reassessment of 2,3,7,8-Tetrachlorodibenzo-*p*-Dioxin (TCDD) and Related Compounds. Part I: Estimating Exposure to Dioxin-Like Compounds. EPA/600/P-00/001Cb. NAS Review Draft. Exposure Assessment and Risk Characterization Group, National Center for Environmental Assessment-Washington Office, Office of Research and Development, U.S. Environmental Protection Agency, Washington, DC. December 2003 [online]. Available: http://www.epa.gov/ncea/pdfs/dioxin/nas-review/#part1 [accessed July 5, 2005].

EPA (U.S. Environmental Protection Agency). 2003b. Exposure and Human Health Reassessment of 2,3,7,8-Tetrachlorodibenzo-*p*-Dioxin (TCDD) and Related Compounds. Part II: Health Assessment of 2,3,7,8-Tetrachlorodibenzo-*p*-Dioxin (TCDD) and Related Compounds. NAS Review Draft. National Center for Environmental Assessment, Office of Research and Development, U.S. Environmental Protection Agency, Washington, DC. December 2003 [online]. Available: http://www.epa.gov/ncea/pdfs/dioxin/nas-review/#part2 [accessed July 5, 2005].

EPA (U.S. Environmental Protection Agency). 2003c. Exposure and Human Health Reassessment of 2,3,7,8-Tetrachlorodibenzo-*p*-Dioxin (TCDD) and Related Compounds. Part III: Integrated Summary and Risk Characterization for 2,3,7,8-Tetrachlorodibenzo-*p*-Dioxin (TCDD) and Related Compounds. NAS Review Draft. National Center for Environmental Assessment, Office of Research and Development, U.S. Environmental Protection Agency, Washington, DC. December 2003 [online]. Available: http://www.epa.gov/ncea/pdfs/dioxin/nas-review/#part3 [accessed July 5, 2005].

EPA (U.S. Environmental Protection Agency). 2003d. Proposed OPPTS Science Policy: PPARα-Mediated Hepatocarcinogenesis in Rodents and Relevance to Human Health Risk Assessments. Office of Prevention, Pesticides and Toxic Substances, U.S. Environmental Protection Agency, Washington, DC [online]. Available: http://www.epa.gov/oscpmont/sap/2003/december9/peroxisomeproliferatorsciencepolicypaper.pdf [accessed July 8, 2005].

EPA (U.S. Environmental Protection Agency). 2004a. An Examination of EPA Risk Assessment Principles and Practices. Staff Paper Prepared for the U.S. Environmental Protection Agency by Members of the Risk Assessment Task Force. EPA/100/B-04/001. Office of the Science Advisor, U.S. Environmental Protection, Washington, DC. March 2004 [online]. Available: http://www.epa.gov/osa/ratf.htm [accessed Sept. 2, 2005].

EPA (U.S. Environmental Protection Agency). 2004b. Toxicological Review of Boron and Compounds. (CAS No. 7440-42-8) in Support of Summary Information on the Integrated Risk Information System (IRIS). EPA 635/04/052. U.S. Environmental Protection Agency, Washington, DC. June 2004 [online]. Available: http://www.epa.gov/iris/toxreviews/0410-tr.pdf [accessed June 24, 2005].

EPA (U.S. Environmental Protection Agency). 2005a. Guidelines for Carcinogen Risk Assessment. EPA/630/P-03/001F. Risk Assessment Forum, U.S. Environmental Protection Agency, Washington, DC. March 2005 [online]. Available: http://www.epa.gov/iriswebp/iris/cancer032505.pdf [accessed July 6, 2005].

EPA (U.S. Environmental Protection Agency). 2005b. The Inventory of Sources and Environmental Releases of Dioxin-Like Compounds in the United States: The Year 2000 Update. EPA/600/P-03/002A. National Center for Environmental Assessment, Washington, DC [online]. Available: http://www.epa.gov/ncea/pdfs/dioxin/2k-update [accessed June 29, 2005].

EPA (U.S. Environmental Protection Agency). 2005c. Air Quality Criteria for Lead (First External Review Draft), Vol. I . EPA/600/R-05/144aA. National Center for Environmental Assessment-RTP Office, Office of Research and Development, U.S. Environmental Protection Agency, Research Triangle Park, NC [online]. Available: http://www.myspy.us/cgi-bin/nph-paidmember.cgi/111011A/http/cfpub.epa.gov/ncea/cfm/recordisplay.cfm?deid=141779 [accessed April 10, 2006].

EPA SAB (U.S. Environmental Protection Agency Science Advisory Board). 2001. Dioxin Reassessment-An SAB Review of the Office of Research and Development's Reassessment of Dioxin. Review of the Revised Sections (Dose Response Modeling, Integrated Summary, Risk Characterization, and Toxicity Equivalency Factors) of the EPA's Reassessment of Dioxin by the Dioxin Reassessment Review Subcommittee of the EPA Science Advisory Board (SAB). EPA-SAB-EC-01-006. Science Advisory Board, Washington, DC. May 2001 [online]. Available: http://www.epa.gov/ttn/atw/ec01006.pdf [accessed Sept. 2, 2005].

Eskenazi, B., P. Mocarelli, M. Warner, S. Samuels, P. Vercellini, D. Olive, L. Needham, D. Patterson, and P. Brambilla. 2000. Seveso Women's Health Study: A study of the effects of 2,3,7,8-tetrachlorodibenzo-*p*-dioxin on reproductive health. Chemosphere 40(9-11): 1247-1253.

Eskenazi, B., P. Mocarelli, M. Warner, S. Samuels, P. Vercellini, D. Olive, L. L. Needham, D.G. Patterson, Jr., P. Brambilla, N. Gavoni, S. Casalini, S. Panazza, W. Turner, and P.M. Gerthoux. 2002a. Serum dioxin concentrations and endometriosis: A cohort study in Seveso, Italy. Environ. Health Perspect. 110(7):629-634.

Eskenazi, B., M. Warner, P. Mocarelli, S. Samuels, L.L. Needham, D.G. Patterson, Jr., S. Lippman, P. Vercellini, P.M. Gerthoux, P. Brambilla, and D. Olive. 2002b. Serum dioxin concentrations and menstrual cycle characteristics. Am. J. Epidemiol. 156(4): 383-392.

Eskenazi, B., P. Mocarelli, M. Warner, W.Y. Chee, P.M. Gerthoux, S. Samuels, L.L. Needham, and D.G. Patterson, Jr. 2003. Maternal serum dioxin levels and birth outcomes in women of Seveso, Italy. Environ. Health Perspect. 111(7):947-953.

Evans, J.S., G.M. Gray, R.L. Sielken, Jr., A.E. Smith, C. Valdez-Flores, and J.D. Graham. 1994a. Use of probabilistic expert judgment in uncertainty analysis of carcinogenic potency. Regul. Toxicol. Pharmacol. 20(1 Pt 1):15-36.

Evans, J.S., J.D. Graham, G.M. Gray, and R.L. Sielken, Jr. 1994b. A distributional approach to characterizing low-dose cancer risk. Risk Anal. 14(1):25-34.

Faustman, E.F., and G.S. Omenn. 2001. Risk assessment. Pp. 83-104 in Casarett and Doull's Toxicology: The Basic Science of Poisons, 6th Ed, C.D. Klaassen, ed. New York: McGraw-Hill.

Farrell, K., L. Safe, and S. Safe. 1987. Synthesis and aryl hydrocarbon receptor binding properties of radiolabeled polychlorinated dibenzofuran congeners. Arch. Biochem. Biophys. 259(1):185-195.

FDA (U.S. Food and Drug Administration). 1997. FDA Requests That Ball Clay Not Be Used in Animal Feeds. CVM Update, October 14, 1997. Center for Veterinary Medicine, U.S. Food and Drug Administration [online]. Available: http://www.fda.gov/cvm/CVM_Updates/ballclay.html [accessed April 4, 2006].

Feeley, M.M. 1995. Workshop on Perinatal Exposure to Dioxin-like Compounds. III. Endocrine effects . Environ. Health Perspect. 103(Suppl. 2):147-150.

Fernandez Salguero, P.M., D.M. Hilbert, S. Rudikoff, J.M. Ward, and F.J. Gonzalez. 1996. Aryl-hydrocarbon receptor-deficient mice are resistant to 2,3,7,8-tetrachlorodibenzo-*p*-dioxin-induced toxicity. Toxicol. Appl. Pharmacol. 140(1):173-179.

Ferrario, J., C. Byrne, M. Lorber, P. Saunders, W. Leese, A. Dupuy, D. Winters, D. Cleverly, J. Schaum, P. Pinsky, C. Deyrup, R. Ellis, and J. Walcott. 1997. A statistical survey of dioxin-like compounds in the United States poultry fat. Organohalogen Compounds 32:245-251.

Fiedler, H., K.R. Cooper, S. Bergek, M. Hjelt, and C. Rappe. 1997. Polychlorinated dibenzo-p-dioxins and polychlorinated dibenzofurans (PCDD/PCDF) in food samples collected in Southern Mississippi, USA. Chemosphere 34(5-7):1411 1419.

Fingerhut, M.A., W.E. Halperin, D.A. Marlow, L.A. Piacitelli, P.A. Honchar, M.H. Sweeney, A.L. Greife, P.A. Dill, K. Steenland, and A.J. Suruda. 1990. Mortality among U.S. Workers Employed in the Production of Chemicals Contaminated with 2,3,7,8-Tetrachlorodibenzo-*p*-dioxin (TCDD). NIOSH/00197158. Industrywide Studies Branch, Division of Surveillance, Hazard Evaluations, and Field Studies, National Institute of Occupational Safety and Health, U.S. Department of Health and Human Services, Cincinnati, OH.

Fingerhut, M.A., W.E. Halperin, D.A. Marlow, L.A. Piacitelli, P.A. Honchar, M.H. Sweeney, A.L. Greife, P.A. Dill, K. Steenland, and A.J. Suruda. 1991. Cancer mortality in workers exposed to 2,3,7,8-tetrachlorodibenzo-*p*-dioxin. N. Engl. J. Med. 324(4):212-218.

Finley, B.L., K.T. Connor, and P.K. Scott. 2003. Use of toxic equivalency factor distributions in probabilistic risk assessments for dioxins, furans and PCBs. J. Toxicol. Environ. Health Part A 66(6):533-550.

Flesch-Janys, D., J. Berger, P. Gurn, A. Manz, S. Nagel, H. Waltsgott, and J.H. Dwyer. 1995. Exposure to polychlorinated dioxins and furans (PCDD/F) and mortality in a cohort of workers from a herbicide-producing plant in Hamburg, Federal Republic of Germany. Am. J. Epidemiol. 142(11):1165-1175.

Flesch-Janys, D., H. Becher, P. Gurn, D. Jung, J. Konietzko, A. Manz, and O. Papke. 1996. Elimination of polychlorinated dibenzo-p-dioxins and dibenzofurans in occupationally exposed persons. J. Toxicol. Environ. Health 47(4):363-378.

Flesch-Janys, D., K. Steindorf, P. Gurn, and H. Becher. 1998. Estimation of the cumulated exposure to polychlorinated dibenzo-*p*-dioxins/furans and standardized mortality ratio analysis of cancer mortality by dose in an occupationally exposed cohort. Environ. Health Perspect. 106(Suppl. 2):655-662.

Fox, T.R., L.L. Best, S.M. Goldsworthy, J.J. Mills, and T.L. Goldsworthy. 1993. Gene expression and cell proliferation in rat liver after 2,3,7,8-tetrachlorodibenzo-*p*-dioxin exposure. Cancer Res. 53(10 Suppl.):2265-2271.

Fries, G.F., and G.S. Marrow. 1975. Retention and excretion of 2,3,7,8-tetrachlorodibenzo-*p*-dioxin by rats. J. Agric. Food Chem. 23(2):265-269.

FSIS (Food Safety and Inspection Service). 2005. Dioxins and Dioxins-Like Compounds in the U.S. Domestic Meat and Poultry Supply. Food Safety and Inspection Service, U.S. Department of Agriculture [online]. Available: http://www.fsis.usda.gov/PDF/Dioxin_Report_0605.pdf. [accessed July 15, 2005].

Gallo, M.A., E.J. Hesse, G.J. MacDonald, T.H. Umbreit. 1986. Interactive effects of estradiol and 2,3,7,8-tetrachlorodibenzo-*p*-dioxin on hepatic cytochrome P-450 and mouse uterus. Toxicol. Lett. 32(1-2):123-132.

Gao, X., D.S. Son, P.F. Terranova, and K.K. Rozman. 1999. Toxic equivalency factors of polychlorinated dibenzo-*p*-dioxins in an ovulation model: Validation of the toxic equivalency concept for one aspect of endocrine disruption. Toxicol. Appl. Pharmacol. 157(2): 107-116.

Gao, X., B.K. Petroff, K.K. Rozman, and P.F. Terranova. 2000a. Gonadotropin-releasing hormone (GnRH) partially reverses the inhibitory effect of 2,3,7,8-tetrachlorodibenzo-*p*-dioxin on ovulation in the immature gonadotropin-treated rat. Toxicology 147(1): 15-22.

Gao, X., P.F. Terranova, and K.K. Rozman. 2000b. Effects of polychlorinated dibenzofurans, biphenyls, and their mixture with dibenzo-*p*-dioxins on ovulation in the gonadotropin-primed immature rat: Support for the toxic equivalency concept. Toxicol. Appl. Pharmacol. 163(2):115-124.

Gao, X., K. Mizuyachi, P.F. Terranova, and K.K. Rozman. 2001. 2,3,7,8-tetrachlorodibenzo-*p*-dioxin decreases responsiveness of the hypothalamus to estradiol as a feedback inducer of preovulatory gonadotropin secretion in the immature gonadotropin-primed rat. Toxicol. Appl. Pharmacol. 170(3):181-190.

Gaylor, D.W., and W. Slikker, Jr. 1990. Risk assessment for neurotoxic effects. Neurotoxicology 11(2):211-218.

Gaylor, D.W., and W. Slikker. 1992. Risk assessment for neurotoxicants. Pp. 331-343 in Neurotoxicology, H. Tilson, and C. Mitchell, eds. New York: Raven Press.

Gehrs, B.C., and R.J. Smialowicz. 1999. Persistent suppression of delayed-type hypersensitivity in adult F344 rats after perinatal exposure to 2,3,7,8-tetrachlorodibenzo-*p*-dioxin. Toxicology 134(1):79-88.

Gerlowski, L.E., and R.K. Jain. 1983. Physiologically based pharmacokinetic modeling: Principles and applications. J. Pharm. Sci. 72(10):1103-1127.

Geyer, H.J., I. Scheuntert, K. Rapp, A. Kettrup, F. Korte, H. Greim, and K. Rozman. 1990. Correlation between acute toxicity of 2,3,7,8-tetrachlorodibenzo-*p*-dioxin (TCDD) and total body fat content in mammals. Toxicology 65(1-2):97-107.

Giavini, E., M. Prati, and C. Vismara. 1983. Embryotoxic effects of 2,3,7,8 tetrachlorodibenzo-*p*-dioxin administered to female rats before mating. Environ. Res. 31(1): 105-110.

Gladen, B.C., N.B. Ragan, and W.J. Rogan. 2000. Pubertal growth and development and prenatal and lactational exposure to polychlorinated biphenyls and dichlorodiphenyl dichloroethene. J. Pediatr. 136(4):490-496.

Goldman, J.M., S.C. Laws, S.K. Balchak, R.L. Cooper, and R.J. Kavlock. 2000. Endocrine-disrupting chemicals: Prepubertal exposures and effects on sexual maturation and thyroid activity in the female rat. A focus on the EDSTAC recommendations. Crit. Rev. Toxicol. 30(2):135-196.

Goodman, D.G., and R.M. Sauer. 1992. Hepatotoxicity and carcinogenicity in female Sprague-Dawley rats treated with 2,3,7,8-tetrachlorodibenzo-*p*-dioxin (TCDD): A pathology working group reevaluation. Regul. Toxicol. Pharmacol. 15(3):245-252.

Grady, A.W., D.L. Fabacher, G. Frame, and B.L. Steadman. 1992. Morphological deformities in brown bullheads administered dietary β-naphthoflavone. J. Aquat. Anim. Health 4(1):7-16.

Graham, M.J., G.W. Lucier, P. Linko, R.R. Maronpot, and J.A. Goldstein. 1988. Increases in cytochrome P-450 mediated 17 beta-estradiol 2-hydroxylase activity in rat liver microsomes after both acute administration and subchronic administration of 2,3,7,8-tetrachlorodibenzo-*p*-dioxin in a two-stage hepatocarcinogenesis model. Carcinogenesis 9(11):1935-1941.

Gray, K.A., M.A. Klebanoff, J.W. Brock, H. Zhou, R. Darden, L. Needham, and M.P. Longnecker. 2005. In utero exposure to background levels of polychlorinated biphenyls and cognitive functioning among school-age children. Am. J. Epidemiol. 162(1):17-26.

Grubbs, W.D., W.H. Wolfe, J.E. Michalek, D.E. Williams, M.B. Lustik, A.S. Brockman, S.C. Henderson, F.R. Burnett, R.J. Land, D.J. Osborne, V.K. Rocconi, M.E. Schreiber, J.C. Miner, G.L. Henriksen, and J.A. Swaby. 1995. Epidemiologic Investigation of Health Effects in Air Force Personnel Following Exposure to Herbicides. AL-TR-920107. Science Applications International Corp., McLean, VA.

Guo, S.W. 2004. The link between exposure to dioxin and endometriosis: A critical reappraisal of primate data. Gynecol. Obstet. Invest. 57(3):157-173.

Guzelian, P.S., M.S. Victoroff, N.C. Halmes, R.C. James, and C.P. Guzelian. 2005. Evidence-based toxicology: A comprehensive framework for causation. Hum. Exp. Toxicol. 24(4):161-201.

Hamm, J.T., C.Y. Chen, and L.S. Birnbaum. 2003. A mixture of dioxins, furans, and non-ortho PCBs based upon consensus toxic equivalency factors produces dioxin-like reproductive effects. Toxicol. Sci. 74(1):182-191.

Harper, N., X. Wang, H. Liu, and S. Safe. 1994. Inhibition of estrogen-induced progesterone receptor in MCF-7 human breast cancer cells by aryl hydrocarbon (Ah) receptor agonists. Mol. Cell Endocrinol. 104(1):47-55.

Harper, P.A., R.D. Prokipcak, L.E. Bush, C.L. Golas, and A.B. Okey. 1991. Detection and characterization of the Ah receptor for 2,3,7,8 tetrachlorodibenzo-*p*-dioxin in the human colon adenocarcinoma cell line LS180. Arch. Biochem. Biophys. 290(1):27-36.

Harper, P.A., J.Y. Wong, M.S. Lam, and A.B. Okey. 2002. Polymorphisms in the human AH receptor. Chem. Biol. Interact. 141(1-2):161-187.

Harrad, S.J., and K.C. Jones. 1992. A source inventory and budget for chlorinated dioxins and furans in the United Kingdom environment. Sci. Total Environ. 126(1-2):89-107.

Harris, M., T. Zacharewski, B. Astroff, and S. Safe. 1989. Partial antagonism of 2,3,7,8-tetrachlorodibenzo-*p*-dioxin-mediated induction of aryl hydrocarbon hydroxylase by 6-methyl-1,3,8-trichlorodibenzofuran: Mechanistic studies. Mol. Pharmacol. 35(5):729-735.

Hayward, D.G., and P.M. Bolger. 2000. PCDD and PCDF levels in baby food made from chicken produced before and after 1997 in the United States. Organohalogen Compounds 47:345-348.

Heimler, I., A.L. Trewin, C.L. Chaffin, R.G. Rawlins, and R.J. Hutz. 1998a. Modulation of ovarian follicle maturation and effects on apoptotic cell death in Holtzman rats exposed to 2,3,7,8-tetrachlorodibenzo-*p*-dioxin (TCDD) in utero and lactationally. Reprod. Toxicol. 12(1):69-73.

Heimler, I., R.G. Rawlins, H. Owen, and R.J. Hutz. 1998b. Dioxin perturbs, in a dose- and time-dependent fashion, steroid secretion, and induces apoptosis of human luteinized granulosa cells. Endocrinology 139(10):4373-4379.

Henriksen, G.L., N.S. Ketchum, J.E. Michalek, and J.A. Swaby. 1997. Serum dioxin and diabetes mellitis in veterans of Operation Ranch Hand. Epidemiology 8(3):252-258.

Henry, E.C., and T.A. Gasiewicz. 1993. Transformation of the aryl hydrocarbon receptor to a DNA-binding form is accompanied by release of the 90 kDa heat-shock protein and increased affinity for 2,3,7,8-tetrachlorodibenzo-*p*-dioxin. Biochem. J. 294(Pt.1):95-101.

Hoff, R.M., D.C.G. Muir, and N.P. Grift. 1992. Annual cycle of polychlorinated biphenyls and organohalogen pesticides in air in southern Ontario: 1. Air concentration data. Environ. Sci. Technol. 26(2)266-275.

Hooiveld, M., D.J. Heederik, and H.B. Bueno-de-Mesquita. 1996. Preliminary results of the second follow-up of a Dutch cohort of workers occupationally exposed to phenoxy herbicides. Organohalogen Compounds 30:185-189.

Hooiveld, M., D.J. Heederik, M. Kogevinas, P. Boffetta, L.L. Needham, D.G. Patterson, Jr., and H.B. Bueno-de-Mesquita. 1998. Second follow-up of a Dutch cohort occupationally exposed to phenoxy herbicides, chlorophenols, and contaminants. Am. J. Epidemiol. 147(9):891-901.

Hunt, G.T., and B.E. Maisel. 1990. Atmospheric PCDDs/PCDFs in wintertime in a northeastern U.S. urban coastal environment. Chemosphere 20(10-12):1455-1462.

Hunt, G.T., B.E. Maisel, and M. Hoyt. 1990. Ambient Concentrations of PCDDs/PCDFs (Polychlorinated Dibenzodioxins/Dibenzofurans) in the South Coast Air Basin. NTIS/PB90-169970. ENSR, Acton, MA.

IARC (International Agency for Research on Cancer). 1997. Polychlorinated Dibenzo para-Dioxins and Polychlorinated Dibenzofurans. IARC Monographs on the Evaluation of Carcinogenic Risks to Humans Vol. 69. Lyon, France: International Agency for Research on Cancer.

Ilsen, A., J.M. Briet, J.G. Koppe, H.J. Pluim, and J. Oosting. 1996. Signs of enhanced neuromotor maturation in children due to perinatal load with background levels of dioxins: Follow-up until age 2 years and 7 months. Chemosphere 33(7):1317-1326.

IOM (Institute of Medicine). 2000. Veterans and Agent Orange: Herbicide/Dioxin Exposure and Type 2 Diabetes. Washington, DC: National Academy Press.

IOM (Institute of Medicine). 2003. Dioxins and Dioxin-like Compounds in the Food Supply: Strategies to Decrease Exposure. Washington, DC: The National Academies Press.

IPCS (International Programme on Chemical Safety). 1994. Assessing Human Health Risks of Chemicals: Derivation of Guidance Values for Health-Based Exposure Limits. Environmental Health Criteria 170. Geneva: World Health Organization [online]. Available: http://www.inchem.org/documents/ehc/ehc/ehc170.htm [accessed June 23, 2005].

IPCS (International Programme on Chemical Safety). 1998a. Polybrominated Dibenzo-*p*-dioxins and Dibenzofurans. Environmental Health Criteria 205. Geneva: World Health Organization [online]. Available: http://www.inchem.org/documents/ehc/ehc/ehc205.htm [accessed June 24, 2005].

IPCS (International Programme on Chemical Safety). 1998b. Executive Summary: Assessment of the Health Risk of Dioxins: Re-evaluation of the Tolerable Daily Intake (TDI). WHO Consultation May 25-29 1998, Geneva, Switzerland. International Programme on Chemical Safety, WHO European Centre for Environment and Health [online]. Available: http://www.who.int/entity/ipcs/ publications/en/exe-sum-final.pdf [accessed Sept. 12, 2005].

IPCS (International Programme on Chemical Safety). 1999. Principles for the Assessment of Risks to Human Health from Exposure to Chemicals. Environmental Health Criteria 210. Geneva: World Health Organization [online]. Available: http://www.inchem.org/documents/ehc/ehc/ehc210.htm [accessed June 23, 2005].

IPCS (International Programme on Chemical Safety). 2004. Announcement of Project for the Re-evaluation of Human and Mammalian Toxic Equivalency Factors (TEFs) of Dioxins and Dioxin-like Compounds and Request for Information, October 2004. World Health Organization [online]. Available: http://www.who.int/ipcs/assessment/tef_review/en/index1.html [accessed June 24, 2005].

Jacobson, J.L., and S.W. Jacobson. 1996. Intellectual impairment in children exposed to polychlorinated biphenyls in utero. N. Engl. J. Med. 335(11):783-789.

Janz, D.M., and G.D. Bellward. 1996. In ovo 2,3,7,8-tetrachlorodibenzo-*p*-dioxin exposure in three avian species. Toxicol. Appl. Pharmacol. 139(2):292-300.

JECFA (Joint FAO/WHO Expert Committee on Food Additives). 2002. Polychlorinated dibenzodioxins, polychlorinated dibenzofurans, and coplanar polychlorinated biphenyls. Pp. 451-658 in Safety Evaluation of Certain Food Additives and Contaminants, WHO Food Additives Series 48. Geneva: World Health Organization [online]. Available: http://www.inchem.org/documents/jecfa/jecmono/v48je20.htm [accessed July 5, 2005].

Jensen, E., and P.M. Bolger. 2001. Exposure assessment of dioxins/furans consumed in dairy foods and fish. Food Addit. Contam. 18(5):395-403.

Jensen, E., R. Canady, and P.M. Bolger. 2000. Exposure assessment for dioxin and furans in seafood and dairy products in the United States, 1998-99. Organohalogen Compounds 47:318-321.

Jeuken, A., B.J. Keser, E. Khan, A. Brouwer, J. Koeman, and M.S. Denison. 2003. Activation of the Ah receptor by extracts of dietary herbal supplements, vegetables and fruits. J. Agric. Food Chem. 51(18):5478-5487.

Jobb, B., M. Uza, R. Hunsinger, K. Roberts, H. Tosine, R. Clement, and B. Bobbie, G. LeBel, D. Williams, and B. Lau. 1990. A survey of drinking water supplies in the Province of Ontario for dioxins and furans. Chemosphere 20(10-12):1553-1558.

Kang, H.K., C.M. Mahan, K.Y. Lee, C.A. Magee, S.H. Mather, and G. Matanoski. 2000. Pregnancy outcomes among U.S. women Vietnam veterans. Am. J. Ind. Med. 38(4): 447-454.

Kerkvliet, N.I., and J.A. Brauner. 1990. Flow cytometric analysis of lymphocyte subpopulations in the spleen and thymus of mice exposed to an acute immunosuppressive dose of 2,3,7,8-tetrachlorodibenzo-*p*-dioxin (TCDD). Environ. Res. 52(2):146-164.

Kerkvliet, N.I., L.B. Steppan, J.A. Brauner, J.A. Deyo, M.C. Henderson, R.S. Tomar, and D.R. Buhler. 1990. Influence of the Ah locus on the humoral immunotoxicity of 2,3,7,8-tetrachlorodibenzo-*p*-dioxin: Evidence for Ah-receptor-dependent and Ah-receptor-independent mechanisms of immunosuppression. Toxicol. Appl. Pharmacol. 105(1):26-36.

Kim, A.H, M.C. Kohn, C.J. Portier, and N.J. Walker. 2002. Impact of physiologically based pharmacokinetic modeling on benchmark dose calculations for TCDD-induced biochemical responses. Regul. Toxicol. Pharmacol. 36(3):287-296.

Kim, D.J., B.S. Han, B. Ahn, R. Hasegawa, T. Shirai, N. Ito, and H. Tsuda. 1997. Enhancement by indole-3-carbinol of liver and thyroid gland neoplastic development in a rat medium-term multiorgan carcinogenesis model. Carcinogenesis 18(2):377-381.

Kociba, R.J., D.G. Keyes, J.E. Beyer, R.M. Carreon, C.E. Wade, D.A. Dittenber, R.P. Kalnins, L.E. Frauson, C.N. Park, S.D. Barnard, R.A. Hummel, and C.G. Humiston. 1978. Results of a two-year chronic toxicity and oncogenicity study of 2,3,7,8-tetrachlorodibenzo-*p*-dioxin in rats. Toxicol. Appl. Pharmacol. 46(2):279-303.

Kociba, R.J., D.G. Keyes, J.E. Beyer, R.M. Carreon, and P.J. Gehring. 1979. Long-term toxicologic studies of 2,3,7,8-tetrachlorodibenzo-*p*-dioxin (TCDD) in laboratory animals. Ann. NY Acad. Sci. 320:397-404.

Kogevinas, M., H. Becher, T. Benn, P.A. Bertazzi, P. Boffetta, H.B. Bueno-de-Mesquita, D. Coggon, D. Colin, D. Flesch-Janys, M. Fingerhut, L. Green, T. Kauppinen, M. Littorin, E. Lynge, J.D. Mathews, M. Neuberger, N. Pearce, and R. Saracci. 1997. Cancer mortality in workers exposed to phenoxy herbicides, chlorophenols, and dioxins. An expanded and updated international cohort study. Am. J. Epidemiol. 145(12):1061-1075.

Kogevinas, M. 2001. Human health effects of dioxins: Cancer, reproductive and endocrine system effects. Hum. Reprod. Update 7(3):331-339.

Kohn, M.C., and R.L. Melnick. 2002. Biochemical origins of the non-monotonic receptor-mediated dose-response. J. Mol. Endocrinol. 29(1):113-123.

Koopman-Esseboom, C., D.C. Morse, N. Weisglas-Kuperus, I.J. Lutkeschipholt, C.G. Van der Paauw, L.G. Tuinstra, A. Brouwer, and P.J. Sauer. 1994. Effects of dioxins and polychlorinated biphenyls on thyroid hormone status of pregnant women and their infants. Pediatr. Res. 36(4):468-473.

Koopman-Esseboom, C., M. Huisman, N. Weisglas-Kuperus, E.R. Boersma, M.A.J. de Ridder, C.G. Van der Paauw, L.G. Tuinstra, and P.J. Sauer. 1995. Dioxin and PCB levels in blood and human milk in relation to living areas in the Netherlands. Chemosphere 29(9-11):2327-2338.

Krishnan, V., and S. Safe. 1993. Polychlorinated biphenyls (PCBs), dibenzo-p-dioxins (PCDDs), and dibenzofurans (PCDFs) as antiestrogens in MCF-7 human breast cancer cells: Quantitative structure-activity relationships. Toxicol. Appl. Pharmacol. 120(1):55-61.

Lahvis, G.P., S.L. Lindell, R.S. Thomas, R.S. McCuskey, C. Murphy, E. Glover, M. Bentz, J. Southard, and C.A. Bradfield. 2000. Portosystemic shunting and persistent fetal vascular structures in aryl hydrocarbon receptor-deficient mice. Proc. Natl. Acad. Sci. USA 97(19):10442-10447.

Lahvis, G.P., R.W. Pyzalski, E. Glover, H.C. Pitot, M.K. McElwee, and C.A. Bradfield. 2005. The aryl hydrocarbon receptor is required for developmental closure of the ductus venosus in the neonatal mouse. Mol. Pharmacol. 67(3):714-720.

Lawson, C.C., T.M. Schnorr, E.A. Whelan, J.A. Deddens, D.A. Dankovic, L.A. Piacitelli, M.H. Sweeney, and L.B. Connally. 2004. Paternal occupational exposure to 2,3,7,8-tetrachlorodibenzo-*p*-dioxin and birth outcomes of offspring: Birth weight, preterm delivery, and birth defects. Environ. Health Perspect. 112(14):1403-1408.

Leibelt, D.A., O.R. Hedstrom, K.A. Fischer, C.B. Pereira, and D.E. Williams. 2003. Evaluation of chronic dietary exposure to indole-3-carbinol and absorption-enhanced 3,3'-diindolylmethane in Sprague-Dawley rats. Toxicol. Sci. 74(1):10-21.

Li, X., D.C. Johnson, and K.K. Rozman. 1995a. Effects of 2,3,7,8-tetrachlorodibenzo-*p*-dioxin (TCDD) on estrous cyclicity and ovulation in female Sprague-Dawley rats. Toxicol. Lett. 78(3):219-222.

Li, X., D.C. Johnson, and K.K. Rozman. 1995b. Reproductive effects of 2,3,7,8-tetrachloro-dibenzo-*p*-dioxin (TCDD) in female rats: Ovulation, hormonal regulation, and possible mechanism(s). Toxicol. Appl. Pharmacol. 133(2):321-327.

Li, X., D.C. Johnson, and K.K. Rozman. 1997. 2,3,7,8-tetrachlorodibenzo-*p*-dioxin (TCDD) increases release of luteinizing hormone and follicle-stimulating hormone from the pituitary of immature female rats in vivo and in vitro. Toxicol. Appl. Pharmacol. 142(2): 264-269.

Lin, T.M., K. Ko, R.W. Moore, D.L. Buchanan, P.S. Cooke, and R.E. Peterson. 2001. Role of the aryl hydrocarbon receptor in the development of control and 2,3,7,8-tetrachloro-dibenzo-*p*-dioxin-exposed male mice. J. Toxicol. Environ. Health Part A 64(4):327-342.

Lorber, M., and L. Phillips. 2002. Infant exposure to dioxin-like compounds in breast milk. Environ Health Perspect. 110(6):A325-A332.

Lorber, M.N., P. Saunders, J. Ferrario, W. Leese, D.L. Winters, D. Cleverly, J. Schaum, C. Deyrup, R. Ellis, J. Walcott, A. Dupuy, C. Byrne, and D. McDanial. 1997. A statistical survey of dioxin-like compounds in the United States pork. Organohalogen Compounds 32:238-244.

Lorber, M.N., D.L. Winters, J. Griggs, R. Cook, S. Baker, J. Ferrario, C. Byrne, A. Dupuy, and J. Schaum. 1998. A national survey of dioxin-like compounds in the United States milk supply. Organohalogen Compounds 38:125-129.

Lucier, G.W., A. Tritscher, T. Goldsworthy, J. Foley, G. Clark, J. Goldstein, and R. Maronpot. 1991. Ovarian hormones enhance 2,3,7,8-tetrachlorodibenzo-*p*-dioxin-mediated increases in cell proliferation and preneoplastic foci in a two-stage model for rat hepatocarcinogenesis. Cancer Res. 51(5):1391-1397.

Lund, A.K., S.L. Peterson, G.S. Timmins, and M.K. Walker. 2005. Endothelin-1-mediated increase in reactive oxygen species and NADPH Oxidase activity in hearts of aryl hydrocarbon receptor (AhR) null mice. Toxicol. Sci. 88(1):265-273.

Lyndon, J.P., F.J. DeMayo, O.M. Conneely, and B.W. O'Malley. 1996. Reproductive phenotypes of the progesterone receptor null mutant mouse. J. Steroid Biochem. Mol. Biol. 56(1-6 Spec.):67-77.

Maisel, B.E., and G.T. Hunt. 1990. Background concentrations of PCDDs/PCDFs in ambient air—A comparison of toxic equivalency factor (TEF) models. Chemosphere 20(7-9): 771-778.

Maronpot, R.R., J.F. Foley, K. Takahashi, T. Goldsworthy, G. Clark, A. Tritscher, C. Portier, and G. Lucier. 1993. Dose response for TCDD promotion of hepatocarcinogenesis in rats initiated with DEN: Histologic, biochemical, and cell proliferation endpoints. Environ. Health Perspect. 101(7):634-642.

Maruyama, W., K. Yoshida, T. Tanaka, and J. Nakanishi. 2002. Possible range of dioxin concentration in human tissues: Simulation with a physiologically based model. Toxicol. Environ. Health. Part A 65(24):2053-2073.

Maruyama, W., K. Yoshida, T. Tanaka, and J. Nakanishi. 2003. Simulation of dioxin accumulation in human tissues and analysis of reproductive risk. Chemosphere 53(4):301-313.

Mathews, T.J., and B.E. Hamilton. 2005. Trend analysis of the sex ratio at birth in the United States. Natl. Vital Stat. Rep. 53(20):1-17.

McKone, T.E., and D.H. Bennett. 2003. Chemical-specific representation of air-soil exchange and soil penetration in regional multimedia models. Environ. Sci. Techn. 37(14):3123-3132.

McKone, T.E., and K.T. Bogen. 1992. Uncertainties in health-risk assessment: An integrated case study based on tetrachloroethylene in California groundwater. Regul. Toxicol. Pharmacol. 15(1):86-103.

Mes, J., and D. Weber. 1989. Non-orthochlorine substituted coplanar polychlorinated biphenyl congeners in Canadian adipose tissue, breast milk and fatty foods. Chemosphere 19(8-9):1357-1365.

Mes, J., W.H. Newsome, and H.B. Conacher. 1991. Levels of specific polychlorinated biphenyl congeners in fatty foods from five Canadian cities between 1986 and 1988. Food Addit. Contam. 8(3):351-361.

Meyer, C., P. O'Keefe, D. Hilker, L. Rafferty, L. Wilson, S. Connor, K. Aldous, K. Markussen, and K. Slade. 1989. A survey of twenty community water systems in New York State for PCDDs and PCDFs. Chemosphere 19(1-6):21-26.

Michalek, J.E., and R.C. Tripathi. 1999. Pharmacokinetics of TCDD in veterans of Operation Ranch Hand: 15-year follow-up. J. Toxicol. Environ. Health A. 57(6):369-378.

Michalek, J.E., R.C. Tripathi, S.P. Caudill, and J.L. Pirkle. 1992. Investigation of TCDD half-life heterogeneity in veterans of Operation Ranch Hand. J. Toxicol. Environ. Health 35(1):29-38.

Mizuyachi, K., D.S. Son, K.K. Rozman, and P.F. Terranova. 2002. Alteration in ovarian gene expression in response to 2,3,7,8-tetrachlorodibenzo-*p*-dioxin: Reduction of cyclooxygenase-2 in the blockage of ovulation. Reprod. Toxicol. 16(3):299-307.

Mocarelli, P., P. Brambilla, P.M. Gerthoux, D.G. Patterson, Jr., and L.L. Needham. 1996. Change in sex ratio with exposure to dioxin [letter]. Lancet 348(9024):409.

Mocarelli, P., P.M. Gerthoux, E. Ferrari, D.G. Patterson, Jr., S.M. Kieszak, P. Brambilla, N. Vincoli, S. Signorini, P. Tramacere, V. Carreri, E.J. Sampson, W.E. Turner, and L.L. Needham. 2000. Paternal concentrations of dioxin and sex ratio of offspring. Lancet 355(9218):1858-1863.

Moolgavkar, S.H., and E.G. Luebeck. 1995. Incorporating cell proliferation kinetics into models for cancer risk assessment. Toxicology 102(1-2):141-147.

MRI (Midwest Research Institute). 1992. Multivariate Statistical Analyses of Dioxin and Furan Levels in Fish, Sediment and Soil Samples Collected Near Resource Recovery Facilities, Final Report. Prepared for Connecticut Department of Environmental Protection, by Midwest Research Institute, Kansas City, MO. December 9, 1992.

Murrell, J.A., C.J. Portier, and R.W. Morris. 1998. Characterizing dose-response: I. Critical assessment of the benchmark dose concept. Risk Anal. 18(1):13-26.

Narasimhan, T.R., A. Craig, L. Arellano, N. Harper, L. Howie, M. Menache, L. Birnbaum, and S. Safe. 1994. Relative sensitivities of 2,3,7,8-tetrachlorodibenzo-*p*-dioxin-induced Cyp1a-1 and Cyp1a-2 gene expression and immunotoxicity in female B6C3F1 mice. Fundam. Appl. Toxicol. 23(4):598-607.

Nazarenko, D.A., S.D. Dertinger, and T.A. Gasiewicz. 2001. In vivo antagonism of AhR-mediated gene induction by 3'-methoxy-4'-nitroflavone in TCDD-responsive lacZ mice. Toxicol. Sci. 61(2):256-264.

NCI (National Cancer Institute). 2002. Levels of Evidence for Human Studies of Cancer Complementary and Alternative Medicine (PDQ(r)). U.S. National Institutes of Health, National Cancer Institute [online]. Available: http://www.nci.nih.gov/cancertopics/pdq/levels-evidence-cam [accessed July 8, 2005].

Needham, L.L., P.M. Gerthoux, D.G. Patterson, Jr., P. Brambilla, J.L. Pirkle, P.L. Tramacere, W.E. Turner, C. Beretta, E.J. Sampson, and P. Mocarelli. 1994. Half-life of 2,3,7,8-tetrachlorodibenzo-*p*-dioxin in serum of Seveso adults: Interim report. Organohalogen Compounds 21:81-85.

Nestrick, T.J., L.L. Lamparski, N.N. Frawley, R.A. Hummel, C.W. Kocher, N.H. Mahle, J.W. McCoy, D.L. Miller, T.L. Peters, J.L. Pillepich, W.E. Smith, and S.W. Tobey. 1986. Perspectives of a large scale environmental survey for chlorinated dioxins: Overview and soil data. Chemosphere 15(9-12):1453-1460.

NIH (National Institutes of Health). 1995. Expert Panel: Report on the Impact and Assessment of Medical and Pathological Waste Incineration on the Bethesda, Maryland, Campus of the National Institutes of Health. Prepared for NIH by EEI, Alexandria, VA.

Nohara, K., H. Izumi, S. Tamura, R. Nagata, and C. Tohyama. 2002. Effect of low-dose 2,3,7,8-tetrachlorodibenzo-*p*-dioxin (TCDD) on influenza A virus-induced mortality in mice. Toxicology 170(1-2):131-138.

NRC (National Research Council). 1983. Risk Assessment in the Federal Government. Washington, DC: National Academy Press.

NRC (National Research Council). 1994. Science and Judgment in Risk Assessment. Washington, DC: National Academy Press.

NRC (National Research Council). 2000. Toxicological Effects of Methylmercury. Washington, DC: National Academy Press.

NTP (National Toxicology Program). 1982a. Carcinogenesis Bioassay of 2,3,7,8-Tetrachlorodibenzo-*p*-Dioxin (CAS No. 1746-01-6) in Osborne-Mendel Rats and B6C3F1 Mice (Gavage Study). NTP Technical Report No. 209. NTP-80-31. NIH 82-1765. U.S. Department of Health and Human Services, Public Health Service, National Institutes of Health, National Toxicology Program, Research Triangle Park, NC, and Bethesda, MD.

NTP (National Toxicology Program). 1982b. Carcinogenesis Bioassay of 2,3,7,8-Tetrachlorodibenzo-*p*-Dioxin (CAS No. 1746-01-6) in Swiss-Webster Mice (Dermal Study). NTP Technical Report No. 201. NTP-80-32. NIH 82-1757. U.S. Department of Health and Human Services, Public Health Service, National Institutes of Health, National Toxicology Program, Research Triangle Park, NC, and Bethesda, MD.

NTP (National Toxicology Program). 2000. 9th Report on Carcinogens. National Toxicology Program, U.S. Department of Health and Human Services, Research Triangle Park, NC.

NTP (National Toxicology Program). 2004. Toxicology and Carcinogenesis Studies of 2,3,7,8-Tetrachlorodibenzo-*p*-dioxin (TCDD) (CAS No. 1746-01-6) in Female Harlan Sprague-Dawley Rats (Gavage Studies). Abstract for TR-521 DRAFT Technical Report, February 17, 2004 [online]. Available: http://ntp server.niehs.nih.gov/index.cfm?objectid=070B69A9-BC89-4234-E4AFAE94C636CC5D [accessed August 19, 2005].

NTP (National Toxicology Program). 2005. Toxicology and Carcinogenesis Studies of a Mixture of 2,3,7,8-Tetrachlorodibenzo-*p*-Dioxin (TCDD) (CAS No. 1746-01-6), 2,3,4,7,8-Pentachlorodibenzofuran (PeCDF) (CAS No. 57117-31-4), and 3,3',4,4',5-Pentachloro-biphenyl (PCB 126) (CAS No. 57465-28-8) in Female Harlan Sprague-Dawley Rats (Gavage Studies). Technical Report No. TR 526. U.S. Department of Health and Human Services, Public Health Service, National Institutes of Health, National Toxicology Program, Research Triangle Park, NC [online]. Available: http://ntp.niehs.nih.gov/index. cfm?objectid=070B7300-0E62-BF12-F4C3E3B5B645A92B [accessed July 8, 2005].

OHEPA (Ohio Environmental Protection Agency). 1995. Dioxin Monitoring Study 1995 Franklin County, Ohio. Division of Air Pollution Control, Ohio Environmental Protection Agency. September 1995.

Oikawa, K., T. Ohbayashi, J. Mimura, R. Iwata, A. Kameta, K. Evine, K. Iwaya, Y. Fujii-Kuriyama, M. Kuroda, and K. Mukai. 2001. Dioxin suppresses the checkpoint protein, MAD2, by an aryl hydrocarbon receptor-independent pathway. Cancer Res. 61(15): 5707-5709.

Ott, M.G., and A. Zober. 1996. Cause specific mortality and cancer incidence among employees exposed to 2,3,7,8-TCDD after a 1953 reactor accident. Occup. Environ. Med. 53(9):606-612.

Oughton, J.A., C.B. Pereira, G.K. DeKrey, J.M. Collier, A.A. Frank, and N.I. Kerkvliet. 1995. Phenotypic analysis of spleen, thymus, and peripheral blood cells in aged C57B1/6 mice following long-term exposure to 2,3,7,8-tetrachlorodibenzo-*p*-dioxin. Fundam. Appl. Toxicol. 25(1):60-69.

Park, J.Y., M.K. Shigenaga, and B.N. Ames. 1996. Induction of cytochrome P4501A1 by 2,3,7,8-tetrachlorodibenzo-*p*-dioxin or indolo(3,2-b)carbazole is associated with oxidative DNA damage. Proc. Natl. Acad. Sci. USA 93(6):2322-2327.

Pavuk, M., A.J. Schecter, F.Z. Akhtar, and J.E. Michalek. 2003. Serum 2,3,7,8-tetrachlorodibenzo-*p*-dioxin (TCDD) levels and thyroid function in Air Force veterans of the Vietnam War. Ann. Epidemiol. 13(5):335-343.

Pearson, R.G., D.L. McLaughlin, and W.D. McIlveen. 1990. Concentrations of PCDD and PCDF in Ontario soils from the vicinity of refuse and sewage sludge incinerators and remote rural and urban locations. Chemosphere 20(10-12):1543-1548.

Pence, B.C., F. Buddingh, and S.P. Yang. 1986. Multiple dietary factors in the enhancement of dimethylhydrazine carcinogenesis: Main effect of indole-3-carbinol. J. Natl. Cancer Inst. 77(1):269-276.

Pesatori, A.C., D. Consonni, S. Bachetti, C. Zocchetti, M. Bonzini, A. Baccarelli, and P.A. Bertazzi. 2003. Short- and long-term morbidity and mortality in the population exposed to dioxin after the "Seveso accident." Ind. Health 41(3):127-138.

Peters, A.K., K. van Londen, A. Bergman, J. Bohonowych, M.S. Denison, M. van den Berg, and J.T. Sanderson. 2004. Effects of polybrominated diphenyl ethers on basal and TCDD-induced ethoxyresorufin activity and cytochrome P450-1A1 expression in MCF-7, HepG2, and H4IIE cells. Toxicol. Sci. 82(2):488-496.

Petroff, B.K., X. Gao, K.K. Rozman, and P.F. Terranova. 2000. Interaction of estradiol and 2,3,7,8-tetrachlorodibenzo-*p*-dioxin (TCDD) in an ovulation model: Evidence for systemic potentiation and local ovarian effects. Reprod. Toxicol. 14(3):247-255.

Petroff, B.K., K.F. Roby, X. Gao, D. Son, S. Williams, D. Johnson, K.K. Rozman, and P.F. Terranova. 2001. A review of mechanisms controlling ovulation with implications for the anovulatory effects of polychlorinated dibenzo-*p*-dioxins in rodents. Toxicology 158(3):91-107.

Petrulis, J.R., and N.J. Bunce. 2000. Competitive behavior in the interactive toxicology of halogenated aromatic compounds. J. Biochem. Mol. Toxicol. 14(2):73-81.

Pluim, H.J., J.J. de Vijlder, K. Olie, J.H. Kok, T. Vulsma, D.A. van Tijn, J.W. van der Slikke, and J.G. Koppe. 1993. Effects of pre- and postnatal exposure to chlorinated dioxins and furans on human neonatal thyroid hormone concentrations. Environ. Health Perspect. 101(6):504-508.

Pohjanvirta, R., M. Unkila, J.T. Tuomisto, O. Vuolteenaho, J. Leppaluoto, and J. Tuomisto. 1993. Effect of 2,3,7,8-tetrachlorodibenzo-*p*-dioxin (TCDD) on plasma and tissue beta-endorphin-like immunoreactivity in the most TCDD-susceptible and the most TCDD-resistant rat strain. Life Sci. 53(19):1479-1487.

Pohjanvirta, R., M. Korkalainen, J. McGuire, U. Simanainen, R. Juvonen, J.T. Tuomisto, M. Unkila, M. Viluksela, J. Bergman, L. Poellinger, and J. Tuomisto. 2002. Comparison of acute toxicities of indolo[3,2-b]carbazole (ICZ) and 2,3,7,8-tetrachlorodibenzo-*p*-dioxin (TCDD) in TCDD-sensitive rats. Food Chem. Toxicol. 40(7):1023-1032.

Poiger, M., and C. Schlatter. 1986. Pharmacokinetics of 2,3,7,8-TCDD in man. Chemosphere 15(9-12):1489-1494.

Poland, A., and E. Glover. 1979. An estimate of the maximum in vivo covalent binding of 2,3,7,8-tetrachlorodibenzo-*p*-dioxin to rat liver protein, ribosomal RNA, and DNA. Cancer Res. 39(9):3341-3344.

Poland, A., D. Palen, and E. Glover. 1982. Tumour promotion by TCDD in skin of HRS/J hairless mice. Nature 300(5889):271-273.

Poland, A., D. Palen, and E. Glover. 1994. Analysis of the four alleles of the murine aryl hydrocarbon receptor. Mol. Pharmacol. 46(5):915-921.

Portier, C., D. Hoel, and J. van Ryzin. 1984. Statistical analysis of the carcinogenesis bioassay data relating to the risks from exposure to 2,3,7,8-tetrachlorodibenzo-*p*-dioxin. Pp. 99-120 in Public Health Risks of the Dioxins, W.W. Lowrance, ed. Los Altos, CA: W. Kaufmann, Inc.

Portier, C.J., and M.C. Kohn. 1996. A biologically-based model for the carcinogenic effects of 2,3,7,8-TCDD in female Sprague-Dawley rats. Organohalogen Compounds 29:222-227.

Portier, C.J., C.D. Sherman, M. Kohn, L. Edler, A. Kopp-Schneider, R.M. Maronpot, and G. Lucier. 1996. Modeling the number and size of hepatic focal lesions following exposure to 2,3,7,8-TCDD. Toxicol. Appl. Pharmacol. 138(1):20-30.

Powell, D.C., R.J. Aulerich, J.C. Meadows, D.E. Tillitt, K.L. Stromborg, T.J. Kubiak, J.P. Giesy, and S.J. Bursian. 1997. Organochlorine contaminants in double-crested cormorants from Green Bay, Wisconsin. II. Effects of an extract derived from cormorant eggs on the chicken embryo. Arch. Environ. Contam. Toxicol. 32(3):316-322.

Prasch, A.L., H. Teraoka, S.A. Carney, W. Dong, T. Hiraga, J.J. Stegeman, W. Heideman, and R.E. Peterson. 2003. Aryl hydrocarbon receptor 2 mediates 2,3,7,8-tetrachlorodibenzo-*p*-dioxin developmental toxicity in zebrafish. Toxicol. Sci. 76(1):138-150.

Puga, A., D.W. Nebert, and F. Carrier. 1992. Dioxin induces expression of c-fos and c-jun proto-oncogenes and a large increase in transcription factor AP-1. DNA Cell Biol. 11(4):269-281.

Puga, A., A. Hoffer, S. Zhou, J.M. Bohm, G.D. Leikauf, and H.G. Shertzer. 1997. Sustained increase in intracellular free calcium and activation of cyclooxygenase-2 expression in mouse hepatoma cells treated with dioxin. Biochem. Pharmacol. 54(12):1287-1296.

Ramadoss, P., and G.H. Perdew. 2004. Use of 2-azido-3-[125I]iodo-7,8-dibromodibenzo-*p*-dioxin as a probe to determine the relative ligand affinity of human versus mouse aryl hydrocarbon receptor in cultured cells. Mol. Pharmacol. 66(1):129-136.

Randerath, K., K.L. Putman, E. Randerath, G. Mason, M. Kelley, and S. Safe. 1988. Organ-specific effects of long term feeding of 2,3,7,8-tetrachlorodibenzo-*p*-dioxin and 1,2,3,7,8-pentachlorodibenzo-*p*-dioxin on I-compounds in hepatic and renal DNA of female Sprague-Dawley rats. Carcinogenesis 9(12):2285-2289.

Rao, M.S., V. Subbarao, J.D. Prasad, and D.G. Scarpelli. 1988. Carcinogenicity of 2,3,7,8-tetrachlorodibenzo-*p*-dioxin in the Syrian golden hamster. Carcinogenesis 9(9):1677-1679.

Rappe, C. 1991. Sources of human exposure to CDDs and PCDFs. Pp. 121-129 in Biological Basis for Risk Assessment of Dioxin and Related Compounds, M.A. Gallo, R.J. Scheuplein, and K.A. van der Heijden, eds. Banbury Report No. 35. Plainview, NY: Cold Spring Harbor Laboratory Press.

Reed, L.W., G.T. Hunt, B.E. Maisel, and M. Hoyt, D. Keefe, and P. Hackney. 1990. Baseline assessment of PCDDs/PCDFs in the vicinity of the Elk River, Minnesota generating station. Chemosphere 21(1-2):159-171.

Revich, B.A. 2002. Chemical substances in the Russian urban environment: Hazard to human health and prospects for its prevention [in Russian]. Vestn. Ross. Akad. Med. Nauk. (9):45-49.

Richards, J.S., T. Jahnsen, L. Hedin, J. Lifka, S. Ratoosh, J.M. Durica, and N.B. Goldring. 1987. Ovarian follicular development: From physiology to molecular biology. Recent. Prog. Horm. Res. 43:231-276.

Roberts, E.A., K.C. Johnson, P.A. Harper, and A.B. Okey. 1990. Characterization of the Ah receptor mediating aryl hydrocarbon hydroxylase induction in the human liver cell line Hep G2. Arch. Biochem. Biophys. 276(2):442-450.

Roby, K.F. 2001. Alterations in follicle development, steroidogenesis, and gonadotropin receptor binding in a model of ovulatory blockade. Endocrinology 142(6):2328-2335.

Roegner, R.H., W.D. Grubbs, M.B. Lustik, A.S. Brockman, and S.C. Henderson. 1991. Air Force Health Study: An Epidemiological Investigation of Health Effects in Air Force Personnel Following Exposure to Herbicides: Serum Dioxin Analysis of 1987 Examination Results. NTIS AD A-237-516. Prepared by Science Applications International Corporation, McLean, VA, for Human Systems Division, Epidemiology Research Division Armstrong Laboratory, Brooks Air Force Base, TX.

Rogowski, D., and W. Yake. 1999. Addendum to Final Report: Screening Survey for Metals and Dioxins in Fertilizer Products and Soils in Washington State. Dioxins in Washington State Agricultural Soils. Ecology Pub. No. 99-333. Environmental Assessment Program, Washington State Department of Ecology, Olympia, WA. November 1999 [online]. Available: http://www.ecy.wa.gov/biblio/99333.html [accessed August 22, 2005].

Rogowski, D., S. Golding, D. Bowhay, and S. Singleton. 1999. Final Report: Screening Survey for Metals and Dioxins in Fertilizer Products and Soils in Washington State. Ecology Pub. No. 99-309. Environmental Assessment Program, Washington State Department of Ecology, Olympia, WA. January 1999 [online]. Available: http://www.ecy.wa.gov/biblio/99309.html [accessed August 23, 2005].

Ross, E.M., and T.P. Kenakin. 2001. Pharmacokinetics: Mechanisms of drug action and the relationship between drug concentration and effect. Pp. 31-43 in Goodman & Gilman's the Pharmacological Basis of Therapeutics, 10th Ed., J.G. Hardman, and L.E. Limbird, eds. New York: McGraw-Hill.

Rozman, K.K., W.L. Roth, H. Greim, B.U. Stahl, and J. Doull. 1993. Relative potency of chlorinated dibenzo-*p*-dioxins (CDDs) in acute, subchronic and chronic (carcinogenicity) toxicity studies: Implications for risk assessment of chemical mixtures. Toxicology 77(1-2):39-50.

Rozman, K.K., B.U. Stahl, L. Kerecsen, and A. Kettrup. 1995. Comparative toxicity of four chlorinated dibenzo-*p*-dioxins (CDDs) and their mixture. IV. Determination of liver concentrations. Arch. Toxicol. 69(8):547-551.

Ryan, J.J., Z. Amirova, and G. Carrier. 2002. Sex ratios of children of Russian pesticide producers exposed to dioxin. Environ. Health Perspect. 110(11):A699-A701.

Rylander, L., U. Stromberg, and L. Hagmar. 1995. Decreased birthweight among infants born to women with a high dietary intake of fish contaminated with persistent organochlorine compounds. Scand. J. Work Environ. Health 21(5):368-375.

Sackett, D.L., S.E. Straus, W.S. Richardson, W. Rosenberg, and R.B. Haynes. 2000. Evidence-Based Medicine: How to Practice and Teach EBM, 2nd Ed. New York: Churchill Livingstone.

Safe, S. 1990. Polychlorinated biphenyls (PCBs), dibenzo-p-dioxins (PCDDs), dibenzofurans (PCDFs), and related compounds: Environmental and mechanistic considerations which support the development of toxic equivalency factors (TEFs). Crit. Rev. Toxicol. 21(1): 51-88.

Sangrujee, N., R.J. Duintjer Tebbens, V.M. Cáceres, and K.M. Thompson. 2003. Policy decision options during the first five years following certification of polio eradication. MedGenMed. 5(4):35.

SCF (Scientific Committee on Food). 2000. Opinion of the SCF on the Risk Assessment of Dioxins and Dioxin-like PCBs in Food, Adapted on 23 November 2000. SCF/CS/CNTM/DIOXIN/8 Final. Scientific Committee on Food, European Commission, Brussel, Belgium [online]. Available: http://europa.eu.int/comm/food/fs/sc/scf/out78_en.pdf [accessed June 24, 2005].

SCF (Scientific Committee on Food). 2001. Opinion of the Scientific Committee on Food on the Risk Assessment of Dioxins and Dioxin-like PCBs in Food, Adapted on 30th May 2001. CS/CNTM/DIOXIN/20 Final. Scientific Committee on Food, European Commission, Brussel, Belgium [online]. Available: http://europa.eu.int/comm/food/fs/sc/scf/out90_en.pdf [accessed June 24, 2005].

Schecter, A., P. Cramer, K. Boggess, J. Stanley, and J.R. Olson. 1997. Levels of dioxins, dibenzofurans, PCB and DDE congeners in pooled food samples collected in 1995 at supermarkets across the United States. Chemosphere 34(5-7):1437-1447.

Schiestl, R.H., J. Aubrecht, W.Y. Yap, S. Kandikonda, and S. Sidhom. 1997. Polychlorinated biphenyls and 2,3,7,8-tetrachlorodibenzo-*p*-dioxin induce intrachromosomal recombination in vitro and in vivo. Cancer Res. 57(19):4378-4383.

Schnorr, T.M., C.C. Lawson, E.A. Whelan, D.A. Dankovic, J.A. Deddens, L.A. Piacitelli, J. Reefhuis, M.H. Sweeney, L.B. Connally, and M.A. Fingerhut. 2001. Spontaneous abortion, sex ratio, and paternal occupational exposure to 2,3,7,8-tetrachlorodibenzo-*p*-dioxin. Environ. Health Perspect. 109(11):1127-1132.

Seidel, S.D., V. Li, G.M. Winter, W.J. Rogers, E.I. Martinez, and M.S. Denison. 2000. Ah receptor-based chemical screening bioassays: Application and limitations for the detection of Ah receptor agonists. Toxicol. Sci. 55(1):107-115.

Shertzer, H.G., D.W. Nebert, A. Puga, M. Ary, D. Sonntag, K. Dixon, L.J. Robinson, E. Cianciolo, and T.P. Dalton. 1998. TCDD causes a sustained oxidative stress response in C57BL/6J mice. Biochem. Biophys. Res. Commun. 253(1):44-48.

Silkworth, J.B., L. Antrim, and L.S. Kaminsky. 1984. Correlations between polychlorinated biphenyl immunotoxicity, the aromatic hydrocarbon locus, and liver microsomal enzyme induction in C57BL/6 and DBA/2 mice. Toxicol. Appl. Pharmacol. 75(1):156-165.

Silkworth, J.B., D.S. Cutler, and G. Sack. 1989. Immunotoxicity of 2,3,7,8-tetrachlorodibenzo-*p*-dioxin in a complex environmental mixture from the Love Canal. Fundam. Appl. Toxicol. 12(2):303-312.

Silkworth, J.B., A. Koganti, K. Illouz, A. Possolo, M. Zhao, and S.B. Hamilton. 2005. Comparison of TCDD and PCB CYP1A induction sensitivities in fresh hepatocytes from human donors, Sprague-Dawley rats, and rhesus monkeys and HepG2 cells. Toxicol. Sci. 87(2):508-519.

Smialowicz, R.J., M.M. Riddle, W.C. Williams, and J.J. Diliberto. 1994. Effects of 2,3,7,8-tetrachlorodibenzo-*p*-dioxin (TCDD) on humoral immunity and lymphocyte subpopulations: Differences between mice and rats. Toxicol. Appl. Pharmacol. 124(2):248-256.

Smith, R.M., P.W. O'Keefe, D.R. Hilker, and K.M. Aldous, S.H. Mo, and R.M. Stelle. 1989. Ambient air and incinerator testing for chlorinated dibenzofurans and dioxins by low resolution mass spectrometry. Chemosphere 18(1-6):585-592.

Smith, R.M., P.W. O'Keefe, K.M. Aldous, H. Valente, S.P. Connor, and R.J. Donnelly. 1990. Chlorinated dibenzofurans and dioxins in atmospheric samples from cities in New York. Environ. Sci. Technol. 24(10):1502-1506.

Smyth, R.L. 2000. Evidence-based medicine. Paediatr. Respir. Rev. 1(3):287-293.

Sommer, R.J., A.J. Hume, J.M. Ciak, J.J. VanNorstrand, M. Friggens, and M.K. Walker. 2005. Early developmental 2,3,7,8-tetrachlorodibenzo-*p*-dioxin exposure decreases chick embryo heart chronotropic response to isoproterenol but not to agents affecting signals downstream of the beta-adrenergic receptor. Toxicol. Sci. 83(2):363-371.

Son, D.S., K. Ushinohama, X. Gao, C.C. Taylor, K.F. Roby, K.K. Rozman, and P.F. Terranova. 1999. 2,3,7,8-tetrachlorodibenzo-*p*-dioxin (TCDD) blocks ovulation by a direct action on the ovary without alteration of ovarian steroidogenesis: Lack of a direct effect on ovarian granulosa and thecal-interstitial cell steroidogenesis in vitro. Reprod. Toxicol. 13(6):521-530.

Staffa, J.A., and M.A. Mehlman, eds. 1979. Innovations in Cancer Risk Assessment (ED01 Study): Proceedings of a Symposium. Park Forest South IL: Pathotox Publishers.

Stahl, B.U., A. Kettrup, and K. Rozman. 1992. Comparative toxicity of four chlorinated dibenzo-*p*-dioxins (CDDs) and their mixture. Part I: Acute toxicity and toxic equivalency factors (TEFs). Arch. Toxicol. 66(7):471-477.

Starr, T.B. 2001. Significant shortcomings of the U.S. Environmental Protection Agency's latest draft risk characterization for dioxin-like compounds. Toxicol. Sci. 64(1):7-13.

Starr, T.B. 2003. Significant issues raised by meta analyses of cancer mortality and dioxin exposure. Environ. Health Perspect. 111(12):1443-1447.

Steenland, K., L. Piacitelli, J. Deddens, M. Fingerhut, and L.I. Chang. 1999. Cancer, heart disease, and diabetes in workers exposed to 2,3,7,8-tetrachlorodibenzo-*p*-dioxin. J. Natl. Cancer Inst. 91(9):779-786.

Steenland, K., J. Deddens, and L. Piacitelli. 2001. Risk assessment for 2,3,7,8-tetrachlorodibenzo-p-dioxin (TCDD) based on an epidemiologic study. Am. J. Epidemiol. 154(5):451-458.

Stohs, S.J., M.A. Shara, N.Z. Alsharif, Z.Z. Wahba, and Z.A. al-Bayati. 1990. 2,3,7,8-Tetrachlorodibenzo-*p*-dioxin-induced oxidative stress in female rats. Toxicol. Appl. Pharmacol. 106(1):126-135.

Suter-Hofmann, M., and C. Schlatter. 1989. Subchronic relay toxicity study with a mixture of polychlorinated dioxins (PCDDs) and polychlorinated furans (PCDFs). Chemosphere 18(1-6):277-282.

ten Tusscher, G.W., P.A. Steerenberg, H. van Loveren, J.G. Vos, A.E. von dem Borne, M. Westra, J.W. van der Slikke, K. Olie, H.J. Pluim, and J.G. Koppe. 2003. Persistent hematologic and immunologic disturbances in 8-year-old Dutch children associated with perinatal dioxin exposure. Environ. Health Perspect. 111(12):1519-1523.

Teraoka, H., W. Dong, Y. Tsujimoto, H. Iwasa, D. Endoh, N. Ueno, J.J. Stegeman, R.E. Peterson, and T. Hiraga. 2003. Induction of cytochrome P450 1A is required for circulation failure and edema by 2,3,7,8-tetrachlorodibenzo-*p*-dioxin in zebrafish. Biochem. Biophys. Res. Commun. 304(2):223-228.

Tewhey Associates. 1997. Letter to Maine Department of Environmental Protection concerning soil sampling data collected from the Yarmouth Pole Yard Site in November 1996. Tewhey Associates, Gorham, ME.

Thackaberry, E.A., Z. Jiang, C.D. Johnson, K.S. Ramos, and M.K. Walker. 2005a. Toxicogenomic profile of 2,3,7,8-tetrachlorodibenzo-*p*-dioxin in the murine fetal heart: Modulation of cell cycle and extracellular matrix genes. Toxicol. Sci. 88(1):231-241.

Thackaberry, E.A., B.A. Nunez, I.D. Ivnitski-Steele, M. Friggins, and M.K. Walker. 2005b. Effect of 2,3,7,8-tetrachlorodibenzo-*p*-dioxin on murine heart development: Alteration in fetal and postnatal cardiac growth, and postnatal cardiac chronotropy. Toxicol. Sci. 88(1):242-249.

Theobald, H.M., and R.E. Peterson. 1997. In utero and lactational exposure to 2,3,7,8-tetrachlorodibenzo-*p*-dioxin: Effects on development of the male and female reproductive system of the mouse. Toxicol. Appl. Pharmacol. 145(1):124-135.

Thompson, K.M., and J.D. Graham. 1996. Going beyond the single number: Using probabilistic risk assessment to improve risk management. Hum. Ecol. Risk Assess. 2(4):1008-1034.

Thompson, K.M., and J.D. Graham. 1997. Producing paper without dioxin pollution. Pp. 203-268 in The Greening of Industry: A Risk Management Approach, J.D. Graham and J.K. Hartwell, eds. Cambridge, MA: Harvard University Press.

Tiernan, T.O., M.L. Taylor, J.H. Garrett, G.F. VanNess, J.G. Solch, D.J. Wagel, G.L. Ferguson, and A. Schecter. 1985. Sources and fate of polychlorinated dibenzodioxins, dibenzofurans and related compounds in human environments. Environ. Health Perspect. 59:145-158.

Tillitt, D.E., and P.J. Wright. 1997. Dioxin-like embryotoxicity of a Lake Michigan lake trout extract to developing lake trout. Organohalogen Compounds 34:221-225.

Toth, K., S. Somfai-Relle, J. Sugar, and J. Bence. 1979. Carcinogenicity testing of herbicide 2,4,5-trichlorophenoxyethanol containing dioxin and of pure dioxin in Swiss mice. Nature 278(5704):548-549.

Tritscher, A.M., A.M. Seacat, J.D. Yager, J.D. Groopman, B.D. Miller, D. Bell, T.R. Sutter, and G.W. Lucier. 1996. Increased oxidative DNA damage in livers of 2,3,7,8-tetrachlorodibenzo-*p*-dioxin treated intact but not ovariectomized rats. Cancer Lett. 98(2):219-225.

Tsafriri, A. 1995. Ovulation as a tissue remodeling process. Proteolysis and cumulus expansion. Adv. Exp. Med. Biol. 377:121-140.

Tsutsumi, O., H. Uechi, H. Sone, J. Yonemoto, Y. Takai, M. Momoeda, C. Tohyama, S. Hashimoto, M. Morita, and Y. Taketani. 1998. Presence of dioxins in human follicular fluid: Their possible stage-specific action on the development of preimplantation mouse embryos. Biochem. Biophys. Res. Commun. 250(2):498-501.

Turteltaub, K.W., J.S. Felton, B.L. Gledhill, J.S. Vogel, J.R. Southon, M.W. Caffee, R.C. Finkel, D.E. Nelson, I.D. Proctor, and J.C. Davis. 1990. Accelerator mass spectrometry in biomedical dosimetry: Relationship between low-level exposure and covalent binding of heterocyclic amine carcinogens to DNA. Proc. Natl. Acad. Sci. USA. 87(14):5288-5292.

Umemura, T., S. Kai, R. Hasegawa, K. Sai, Y. Kurokawa, and G.M. Williams. 1999. Pentachlorophenol (PCP) produces liver oxidative stress and promotes but does not initiate hepatocarcinogenesis in B6C3F1 mice. Carcinogenesis 20(6):1115-1120.

Valdez, K.E., and B.K. Petroff. 2004. Potential roles of the aryl hydrocarbon receptor in female reproductive senescence. Reprod. Biol. 4(3):243-258.

Van Birgelen, A.P., J. Van der Kolk, K.M. Fase, I. Bol, H. Poiger, A. Brouwer, and M. Van den Berg. 1994a. Toxic potency of 3,3',4,4',5-pentachlorobiphenyl relative to and in combination with 2,3,7,8-tetrachlorodibenzo-*p*-dioxin in a subchronic feeding study in the rat. Toxicol. Appl. Pharmacol. 127(2):209-221.

Van Birgelen, A.P., J. Van der Kolk, K.M. Fase, I. Bol, H. Poiger, M. Van den Berg, and A. Brouwer. 1994b. Toxic potency of 2,3,3',4,4',5-hexachlorobiphenyl relative to and in combination with 2,3,7,8-tetrachlorodibenzo-*p*-dioxin in a subchronic feeding study in the rat. Toxicol. Appl. Pharmacol. 126(2):202-213.

Van Birgelen, A.P., K.M. Fase, J. van der Kolk, H. Poiger, A. Brouwer, W. Seinen, and M. van den Berg. 1996. Synergistic effect of 2,2',4,4',5,5'-hexachlorobiphenyl and 2,3,7,8-tetrachlorodibenzo-*p*-dioxin on hepatic porphyrin levels in the rat. Environ. Health Perspect. 104(5):550-557.

Van den Berg, M., K. Olie, and O. Hutzinger. 1985. Polychlorinated dibenzofurans (PCDFs): Environmental occurrence, physical, chemical and biological properties. Toxicol. Environ. Chem. 9(3):171-217.

Van den Berg, M., J. De Jongh, H. Poiger, and J.R. Olson. 1994. The toxicokinetics and metabolism of polychlorinated dibenzo-*p*-dioxins (PCDDs) and dibenzofurans (PCDFs) and their relevance for toxicity. Crit. Rev. Toxicol. 24(1):1-74.

Van den Berg, M., L. Birnbaum, A.T. Bosveld, B. Brunstrom, P. Cook, M. Feeley, J.P. Giesy, A. Hanberg, R. Hasegawa, S.W. Kennedy, T. Kubiak, J.C. Larsen, F.X. van Leeuwen, A.K. Liem, C. Nolt, R.E. Peterson, L. Poellinger, S. Safe, D. Schrenk, D. Tillitt, M. Tysklind, M. Younes, F. Waern, and T. Zacharewski. 1998. Toxic equivalency factors (TEFs) for PCBs, PCDDs, PCDFs for humans and wildlife. Environ. Health Perspect. 106(12):775-792.

Van Den Heuvel, R.L., G. Koppen, J.A. Staessen, E.D. Hond, G. Verheyen, T.S. Nawrot, H.A. Roels, R. Vlietinck, and G.E. Schoeters. 2002. Immunologic biomarkers in relation to exposure markers of PCBs and dioxins in Flemish adolescents (Belgium). Environ. Health Perspect. 110(6):595-600.

Van Oostdam, J.C., and J.E.H. Ward. 1995. Dioxins and Furans in the British Columbia Environment. Victoria: BC Environment, Environmental Protection Department.

Vasquez, A., N. Atallah-Yunes, F.C. Smith, X. You, S.E. Chase, A.E. Silverstone, and K.L. Vikstrom. 2003. A role for the aryl hydrocarbon receptor in cardiac physiology and function as demonstrated by AhR knock out mice. Cardiovasc. Toxicol. 3(2):153-163.

Vecchi, A., A. Mantovani, M. Sironi, W. Luini, M. Cairo, and S. Garattini. 1980. Effect of acute exposure to 2,3,7,8-tetrachlorodibenzo-*p*-dioxin on humoral antibody production in mice. Chem. Biol. Interact. 30(3):337-342.

Vecchi, A., M. Sironi, M.A. Canegrati, M. Recchia, and S. Garattini. 1983. Immunosuppressive effects of 2,3,7,8-tetrachlorodibenzo-*p*-dioxin in strains of mice with different susceptibility to induction of aryl hydrocarbon hydroxylase. Toxicol. Appl. Pharmacol. 68(3):434-441.

Versar, Inc. 1996. Results of Survey Design for Measuring Dioxin-Like Compounds in Edible Vegetable Oils. Draft report. EPA Contract No. 68-D3-0013. Prepared by Versar, Inc., for U.S. Environmental Protection Agency, National Center for Environmental Assessment.

Viluksela, M., B.U. Stahl, L.S. Birnbaum, K.W. Schramm, A. Kettrup, and K.K. Rozman. 1997a. Subchronic/chronic toxicity of 1,2,3,4,6,7,8-heptachlorodibenzo-*p*-dioxin (HpCDD) in rats. Part I. Design, general observations, hematology, and liver concentrations. Toxicol. Appl. Pharmacol. 146(2):207-216.

Viluksela, M., B.U. Stahl, L.S. Birnbaum, and K.K. Rozman. 1997b. Subchronic/chronic toxicity of 1,2,3,4,6,7,8-heptachlorodibenzo-*p*-dioxin (HpCDD) in rats. Part II. Biochemical effects. Toxicol. Appl. Pharmacol. 146(2):217-226.

Viluksela, M., B.U. Stahl, L.S. Birnbaum, and K.K. Rozman. 1998a. Subchronic/chronic toxicity of a mixture of four chlorinated dibenzo-*p*-dioxins in rats. II. Biochemical effects. Toxicol. Appl. Pharmacol. 151(1):70-78.

Viluksela, M., B.U. Stahl, L.S. Birnbaum, K.W. Schramm, A. Kettrup, and K.K. Rozman. 1998b. Subchronic/chronic toxicity of a mixture of four chlorinated dibenzo-*p*-dioxins in rats. I. Design, general observations, hematology and liver concentrations. Toxicol. Appl. Pharmacol. 151(1):57-69.

Viluksela, M., Y. Bager, J.T. Tuomisto, G. Scheu, M. Unkila, R. Pohjanvirta, S. Flodstrom, V.M. Kosma, J. Maki-Paakkanen, T. Vartiainen, C. Klimm, K.W. Schramm, L. Warngard, and J. Tuomisto. 2000. Liver tumor-promoting activity of 2,3,7,8-tetrachlorodibenzo-*p*-dioxin (TCDD) in TCDD-sensitive and TCDD-resistant rat strains. Cancer Res. 60(24):6911-6920.

Vreugdenhil, H.J., C.I. Lanting, P.G. Mulder, E.R. Boersma, and N. Weisglas-Kuperus. 2002a. Effects of prenatal PCB and dioxin background exposure on cognitive and motor abilities in Dutch children at school age. J. Pediatr. 140(1):48-56.

Vreugdenhil, H.J., F.M. Slijper, P.G. Mulder, and N. Weisglas-Kuperus. 2002b. Effects of perinatal exposure to PCBs and dioxins on play behavior in Dutch children at school age. Environ. Health Perspect. 110(10):A593-A598.

Walisser, J.A., M.K. Bunger, E. Glover, and C. Bradfield. 2004a. Gestational exposure of Ahr and Arnt hypomorphs to dioxin rescues vascular development. Proc. Natl. Acad. Sci. USA 101(47):16677-16682.

Walisser, J.A., M.K. Bunger, E. Glover, E.B. Harstad, and C.A. Bradfield. 2004b. Patent ductus venosus and dioxin resistance in mice harboring a hypomorphic Arnt allele. J. Biol. Chem. 279(16):16326-16331.

Walker, N.J., P.W. Crockett, A. Nyska, A.E. Brix, M.P. Jokinen, D.M. Sells, J.R. Hailey, M. Easterling, J.K. Haseman, M. Yin, M.E. Wyde, J.R. Bucher, and C.J. Portier. 2005. Dose-additive carcinogenicity of a defined mixture of "dioxin-like compounds." Environ. Health Perspect. 113(1):43-48.

Wang, X., M.J. Santostefano, M.V. Evans, V.M. Richardson, J.J. Diliberto, and L.S. Birnbaum. 1997. Determination of parameters responsible for pharmacokinetic behavior of TCDD in female Sprague-Dawley rats. Toxicol. Appl. Pharmacol. 147(1): 151-168.

Warner, M., S. Samuels, P. Mocarelli, P.M. Gerthoux, L. Needham, D.G. Patterson, Jr., and B. Eskenazi. 2004. Serum dioxin concentrations and age at menarche. Environ. Health Perspect. 112(13):1289-1292.

Warren, T.K., K.A. Mitchell, B.P. Lawrence. 2000. Exposure to 2,3,3,7,8 tetracholrodibenzo-*p*-dioxin (TCDD) suppresses the humoral and cell-mediated immune response to influenza A virus without affecting cytolytic activity in the lung. Toxicol. Sci. 56(1):114-123.

Wassenberg, D.M., and R.T. Di Giulio. 2004a. Synergistic embryotoxicity of polycyclic aromatic hydrocarbon aryl hydrocarbon receptor agonists with cytochrome P4501A inhibitors in Fundulus heteroclitus. Environ. Health Perspect. 112(17):1658-1664.

Wassenberg, D.M., and R.T. Di Giulio. 2004b. Teratogenesis in Fundulus heteroclitus embryos exposed to a creosote-contaminated sediment extract and CYP1A inhibitors. Mar. Environ. Res. 58(2-5):163-168.

Weber, L.W., M. Lebofsky, B.U. Stahl, A. Kettrup, and K. Rozman. 1992. Comparative toxicity of four chlorinated dibenzo-p-dioxins (CDDs) and their mixture. Part II: Structure-activity relationships with inhibition of hepatic phosphoenolpyruvate carboxykinase, pyruvate carboxylase, and gamma-glutamyl transpeptidase activities. Arch. Toxicol. 66(7):478-483.

Weber, L.W., C.D. Palmer, and K. Rozman. 1994. Reduced activity of tryptophan 2,3-dioxygenase in the liver of rats treated with chlorinated dibenzo-*p*-dioxins (CDDs): Dose-responses and structure-activity relationship. Toxicology 86(1-2):63-69.

Weisglas-Kuperus, N., T.C. Sas, C. Koopman-Esseboom, C.W. van der Zwan, M.A. De Ridder, A. Beishuizen, H. Hooijkaas, and P.J. Sauer. 1995. Immunologic effects of background prenatal and postnatal exposure to dioxins and polychlorinated biphenyls in Dutch infants. Pediatr. Res. 38(3):404-410.

Weisglas-Kuperus, N., S. Patandin, G.A. Berbers, T.C. Sas, P.G. Mulder, P.J. Sauer, and H. Hooijkaas. 2000. Immunologic effects of background exposure to polychlorinated biphenyls and dioxins in Dutch preschool children. Environ. Health Perspect. 108(12): 1203-1207.

Weisglas-Kuperus, N., H.J. Vreugdenhil, and P.G. Mulder. 2004. Immunological effects of environmental exposure to polychlorinated biphenyls and dioxins in Dutch school children. Toxicol. Lett. 149(1-3):281-285.

Whitlock, J.P. Jr. 1989. The control of cytochrome P-450 gene expression by dioxin. Trends Pharmacol Sci. 10(7):285-288.

WHO (World Health Organization). 1992. International Statistical Classification of Diseases and Related Health Problems: Tenth Revision (ICD-10). Geneva: World Health Organization.

WHO (World Health Organization). 2005. Chemical-Specific Adjustment Factors for Interspecies Differences and Human Variability: Guidance Document for Use of Data in Dose/Concentration-Response Assessment. Harmonization Project Document No. 2. Geneva: World Health Organization [online]. Available: http://whqlibdoc.who.int/publications/2005/9241546786_eng.pdf [accessed April 11, 2006].

Whysner, J., and G.M. Williams. 1996. 2,3,7,8-Tetrachlorodibenzo-*p*-dioxin mechanistic data and risk assessment: Gene regulation, cytotoxicity, enhanced cell proliferation, and tumor promotion. Pharmacol. Ther. 71(1-2):193-223.

Wilker, C., L. Johnson, and S. Safe. 1996. Effects of developmental exposure to indole-3-carbinol or 2,3,7,8-tetrachlorodibenzo-*p*-dioxin on reproductive potential of male rat offspring. Toxicol. Appl. Pharmacol. 141(1):68-75.

Williams, G.M. 1992. DNA reactive and epigenetic carcinogens. Exp. Toxicol. Pathol. 44(8): 457-463.

Williams, G.M. 1997. Chemicals with carcinogenic activity in the rodent liver; mechanistic evaluation of human risk. Cancer Lett. 117(2):175-188.

Williams, G.M., and M.J. Iatropoulos. 2002. Alteration of liver cell function and proliferation: Differentiation between adaptation and toxicity. Toxicol. Pathol. 30(1):41-53.

Williams, G.M., M.J. Iatropoulos, and A.M. Jeffrey. 2005. Thresholds for DNA-reactive (genotoxic) organic carcinogens. J. Toxicol. Pathol. 18(1):69-77.

Williams, S.N., H. Shih, D.K. Guenette, W. Brackney, M.S. Denison, G.V. Pickwell, and L.C. Quattrochi. 2000. Comparative studies on the effects of green tea extracts and individual tea catechins on human CYP1A gene expression. Chem. Biol. Interact. 128(3):211-229.

Winters, D., D. Cleverly, K. Meier, A. Dupuy, C. Byrne, C. Deyrup, R. Ellis, J. Ferrario, R. Harless, W. Leese, M. Lorber, D. McDaniel, J. Schaum, and J. Walcott. 1996a. A statistical survey of dioxin-like compounds in United States beef: A progress report. Chemosphere 32(3):469-478.

Winters, D., D. Cleverly, M. Lorber, K. Meier, A. Dupuy, C. Byrne, C. Deyrup, R. Ellis, J. Ferrario, W. Leese, J. Schaum, and J. Wolcott. 1996b. Coplanar polychlorinated biphenyls (PCBs) in national sample of beef in the United States: Preliminary results. Organohalogen Compounds 23:350-354 [online]. Available: http://www.epa.gov/ncea/pdfs/pcbbeef.pdf [accessed August 24, 2005].

Wolfe, W.H., J.E. Michalek, J.C. Miner, J.L. Pirkle, S.P. Caudill, D.G. Patterson Jr., and L.L. Needham. 1994. Determinants of TCDD half-life in veterans of operation ranch hand. J. Toxicol. Environ. Health 41(4):481-488.

Wyde, M.E., V.A. Wong, A.H. Kim, G.W. Lucier, and N.J. Walker. 2001. Induction of hepatic 8-oxo-deoxyguanosine adducts by 2,3,7,8-tetrachlorodibenzo-*p*-dioxin in Sprague-Dawley rats is female-specific and estrogen-dependent. Chem. Res. Toxicol. 14(7):849-855.

Xu, L., A.P. Li, D.L. Kaminski, and M.F. Ruh. 2000. 2,3,7,8 Tetrachlorodibenzo-*p*-dioxin induction of cytochrome P4501A in cultured rat and human hepatocytes. Chem. Biol. Interact. 124(3):173-189.

Yao, C., and S. Safe. 1989. 2, 3,7,8-Tetrachlorodibenzo-*p*-dioxin-induced porphyria in genetically inbred mice: Partial antagonism and mechanistic studies. Toxicol. Appl. Pharmacol. 100(2):208-216.

Yoshida, M., S. Katashima, J. Ando, T. Tanaka, F. Uematsu, D. Nakae, and A. Maekawa. 2004. Dietary indole-3-carbinol promotes endometrial adenocarcinoma development in rats initiated with *N*-ethyl-*N*′-nitro-*N*-nitrosoguanidine, with induction of cytochrome P450s in the liver and consequent modulation of estrogen metabolism. Carcinogenesis 25(11):2257-2264.

Yu, M.L., J.W. Hsin, C.C. Hsu, W.C. Chan, and Y.L. Guo. 1995. The immunologic evaluation of the Yucheng children. Chemosphere 37(9-12):1855-1865.

Zhang, S., C. Qin, and S. Safe. 2003. Flavonoids as aryl hydrocarbon receptor agonists/antagonists: Effects of structure and cell context. Environ. Health Perspect. 111(16):1877-1882.

Appendixes

A

Biographical Information on Committee Members

From left to right: Nancy Kim, Alvaro Puga, Malcolm Pike, Michael Denison, Andrew Renwick, Thomas McKone, Richard Di Giulio, David Savitz, David Eaton, Norbert Kaminski, Joshua Cohen, Paul Terranova, Allen Silverstone, Gary Williams, Yiliang Zhu, Kimberly Thompson, Dennis Bier

David L. Eaton, *Chair*, is a professor in the Department of Environmental and Occupational Health Sciences and the Public Health Genetics Program in the school of Public Health and Community Medicine, and associate vice provost for research at the University of Washington in Seattle. He is also the director of the Center of Ecogenetics and Environmental Health at the university and an associate director of the Fred Hutchinson Cancer Research Center–University of Washington–Childrens' Hospital and Medical Center Cancer Center Research Consortium. He earned a B.S. in premedicine from Montana State University in 1974 and a Ph.D. in pharmacology and toxicology from the University of Kansas Medical Center in 1978. Dr. Eaton's research interests include the molecular basis of chemically induced cancers and understanding how human genetic variation in biotransformation enzymes may increase or decrease individual susceptibility to natural and synthetic chemicals found in the environment. He has served on numerous boards and committees, including service as president of the Society of Toxicology in 2001-2002 and as a member of the NRC Board on Environmental Studies and Toxicology (BEST). Dr. Eaton has served as chair of the NRC Committee on Emerging Issues and Data on Environmental Contaminants and as a member of the Panel on Arsenic in Drinking Water. Dr. Eaton has been awarded many distinguished fellowships and honors, including the Achievement Award from the Society of Toxicology in 1990. He is an elected fellow of the Academy of Toxicological Sciences and the American Association for the Advancement of Science.

Dennis M. Bier is professor of pediatrics and director of the Children's Nutrition Research Center and program director of the General Clinical Research Center at Baylor College of Medicine. Dr. Bier earned a B.S. from Le Moyne College in 1962 and an M.D. from New Jersey College of Medicine in 1966. Dr. Bier's research interests include the role of nutrition in human health and in the prevention and treatment of disease and the role of maternal, fetal, and childhood nutrition on the growth, development, and health of children through adolescence. He also has professional interests in the long-term consequences of nutrient inadequacy during critical periods of embryonic and fetal life through infancy and childhood and on the pathogenesis of adult chronic diseases. Dr. Bier has expertise in macronutrients (carbohydrate, lipid, and protein), intermediary metabolism, tracer kinetics, diabetes, obesity, and endocrine disorders. Dr. Bier was elected to the Institute of Medicine in 1997 and was a member of IOM's Food and Nutrition Board. He also served on the IOM Committee on Implications of Dioxin in the Food Supply.

Joshua T. Cohen is a lecturer at Tufts New England Medical Center in the Institute for Clinical Care Research and Health Policy Studies. He earned his B.A. (1986) in applied mathematics and Ph.D. (1994) in decision sciences from Harvard University. Dr. Cohen's research focuses on the application of decision analytical techniques to environmental risk management problems with a special emphasis on the proper characterization and analysis of uncertainty. He was the lead author on a study comparing the risks and benefits of changes in population fish consumption patterns, an analysis of the risks and benefits of cell-phone use while driving, and a study comparing the costs and health impacts of advanced diesel and compressed natural gas urban transit buses. He has also played a key role in a risk assessment of bovine spongiform encephalopathy ("mad cow disease") in the United States.

Michael Denison is a professor in the Department of Environmental Toxicology at the University of California at Davis. He earned a B.S. from Saint Francis College in 1977, an M.S. from Mississippi State University in 1980, and a Ph.D. from Cornell University in 1983. Dr. Denison completed postdoctoral training at the Hospital for Sick Children in Toronto, Canada, and the Department of Pharmacology at Stanford University. He began his professional career as an assistant professor in the Department of Biochemistry at Michigan State University in 1988 and relocated to the University of California in 1992. His research interests include the biochemical and molecular mechanisms by which xenobiotics (particularly dioxins and related chemicals and endocrine disruptors) interact with ligand-dependent transcription factors to produce biological and toxicological effects in animals. Dr. Denison is also examining the molecular and structural characteristics of the Ah receptor responsible for its binding and activation by dioxins and structurally diverse xenobiotics. The application of molecular biological approaches for the development of rapid high-throughput bioassay systems for detection and characterization of ligands for xenobiotic receptors present in environmental, biological, and food samples is another major research area. Dr. Denison is co-chair of the International Advisory Board of the annual International Dioxin Symposium and was the organizer and co-chair of the 2003 U.S.-Vietnam Scientific Workshop on Methodologies of Dioxin Screening, Remediation, and Site Characterization in Hanoi, Vietnam.

Richard Di Giulio is a professor of environmental toxicology at the Nicholas School of the Environment and Earth Sciences, director of the Integrated Toxicology Program, director of the Superfund Basic Research Center, and director of the Center for Comparative Biology of Vulnerable Populations at Duke University. He earned a B.A. from the University of Texas at Austin in 1972, an M.S. from Louisiana State University in 1978, and a Ph.D. from Virginia Polytechnic Institute and State University in 1982. Dr. Di Giulio's professional experience began as an assistant professor and research associate at the School of Forestry and Environmental Studies at Duke University in 1982. His research focuses on biochemical and molecular responses of aquatic animals to environmental stressors, particularly contaminants. Of particular concern are mechanisms of oxidative metabolism of aromatic hydrocarbons; mechanisms of free radical production and antioxidant defense, and mechanisms of chemical carcinogenesis, developmental perturbations, and adaptations to contaminated environments by fishes. Dr. Di Giulio also has interests in the area of interconnections between ecological and human health.

Norbert Kaminski is the director of the Center for Integrative Toxicology, formerly known as the Institute for Environmental Toxicology at Michigan State University. He is also a professor of pharmacology and toxicology in the Department of Pharmacology and Toxicology. Dr. Kaminski earned a B.A. in chemistry from Loyola University in 1978, an M.S. in toxicology in 1981, and a Ph.D. in toxicology and physiology in 1985 from North Carolina State University. Dr. Kaminski's postdoctoral training was in immunotoxicology at the Medical College of Virginia. He continued at the Medical College of Virginia as a faculty member until 1993. His research interests are in the areas of immunotoxicology and immunopharmacology and, in particular, the molecular mechanisms by which dioxins alter B-cell differentiation and function. Dr. Kaminski served on the IOM Committee to Review the Health Effects in Vietnam Veterans of Exposure to Herbicides.

Nancy Kim is the director of the Division of Environmental Health Assessment, within the New York State Department of Health, and an associate professor at the University of Albany School of Public Health. She earned a B.A. in chemistry from the University of Delaware in 1964, and an M.S. and Ph.D. in chemistry from Northwestern University in 1966 and 1969, respectively. Her interests include toxicological evaluations, exposure assessments, risk assessment, structural activity correlations, and quantitative relationships between toxicological parameters. Dr. Kim has held numerous panel memberships and is now a member of the National Center for Environmental Health/Agency for Toxic Substances and Disease Registry Board of Scientific Counselors. She has received several awards and honors, including the Women in Government Award presented by the New York State Department of Health. In 1999, Dr. Kim was inducted into the Delta Omega Society, a national honorary public health society.

Antoine Keng Djien Liem is scientific coordinator of the Scientific Committee of the European Food Safety Authority (EFSA) in Parma. Dr. Liem earned an M.Sc. degree in environmental chemistry and toxicology from the University of Amsterdam (1984) and a Ph.D. in biology from the Utrecht University (1997). Following his university study in Amsterdam, Dr. Liem began his career at the Department of Industrial Contaminants of the Dutch National Institute of Public Health and the Environment (RIVM) in Bilthoven. After the discovery of increased levels of dioxins in milk in cows grazing in the vicinity of municipal waste incinerators, Dr. Liem was leader of various multidisciplinary dioxin projects. He was appointed chairman of the Dutch Working Group on Dioxins in Food and the Dutch Working

Group on Dietary Intakes and acted as temporary adviser and national delegate in the framework of studies of the World Health Organization related to dioxins and related compounds. In 1998-2000, Dr. Liem acted as the leader of the Dutch delegation coordinating a European Scientific Cooperation (SCOOP) Task on the assessment of dietary intake of dioxins and related PCBs by the population of EU member states. This EU task was jointly coordinated by RIVM and the Swedish National Food Administration. The outcomes of the EU-SCOOP project were used in the risk assessments of dioxins and dioxin-like compounds in food carried out by the EU's Scientific Committee on Food (SCF), in which Dr. Liem contributed to the Task Force preparing the opinion, and the Joint FAO/WHO Expert Committee on Food Additives in 2001.

Thomas McKone is senior staff scientist and deputy department head at the Lawrence Berkeley National Laboratory and an adjunct professor and researcher at the University of California at Berkeley School of Public Health. He earned a Ph.D. from the University of California at Los Angeles in 1981. Dr. McKone's research interests include risk assessment methods, mass transfer at environmental and human-environmental boundaries, model uncertainty and reliability in exposure risk assessment, environmental and occupational radioactivity, and biotransfer and bioconcentration. He is very active in many research and professional organizations and is a member of the NRC Committee on the Selection and Use of Models in the Regulatory Decision Process and was a member of the NRC Committee on Toxicants and Pathogens in Biosolids Applied to Land. Dr. McKone is also a member of the Advisory Council of the American Center for Life Cycle Assessment and a member of the Organizing Committee for the International Life-Cycle Initiative, a joint effort of the United Nations Environment Program and the Society for Environmental Toxicology and Chemistry. One of Dr. McKone's most recognized achievements was his development of the CalTOX risk assessment framework for the California Department of Toxic Substances Control.

Malcolm Pike is a professor in the Department of Preventive Medicine at the Norris Comprehensive Cancer Center at the University of Southern California Keck School of Medicine. As a native of South Africa, he earned a B.S. (honors) in mathematics from the University of Witwaterstand in Johannesburg, South Africa, in 1956. He then studied statistics at Birkbeck College of the University of London and earned a diploma in mathematical statistics from Cambridge University in 1958. Dr. Pike received a Ph.D. in mathematical statistics from Aberdeen University in Aberdeen, Scotland, in 1963. From 1963 to 1969, he was at the Statistical Research Unit of the Medical Research Council at University College, London, and from 1969 through 1973, he was the Regius Professor of Medicine at Oxford University. Between 1973 and 1983, he held the position of professor of preventive medicine at the University of Southern California School of Medicine. Following this, Dr. Pike was the director of the ICRF Cancer Epidemiology and Clinical Trials Unit at Oxford University for 4 years. His research areas include the epidemiology of breast, endometrial, and ovarian cancer. Dr. Pike has received many distinguished honors, including the Brinker International Award of the Susan G. Komen Breast Cancer Foundation in 1994 and the American Association for Cancer Research Award for Research Excellence in Cancer Epidemiology and Prevention in 2004. He was elected to the Institute of Medicine in 1994.

Alvaro Puga is a professor of molecular biology and environmental health in the Department of Environmental Health at the University of Cincinnati, director of the Center for Environmental Genetics and deputy director of the Superfund Basic Research Program at the University of Cincinnati. In Spain, he earned a Licenciate in Biology degree in 1966 from the Universidad Complutense in Madrid. In the United States, Dr. Puga earned a Ph.D in 1972 from Purdue University and completed his postdoctoral training in 1976 with Scripps Clinic and Research Foundation in La Jolla, California. His research interests include the molecular mechanisms of dioxin and other environmental contaminants with the purpose of elucidating the signal transduction pathways that underlie the biological responses postexposure to these contaminants. He is also investigating the genetic diversity on the response to exposure, specifically the genes that code for transcription factors with a regulatory role in the expression of detoxification enzymes. Before joining the University of Cincinnati, he was the head of the Unit on Pharmacogenetics, Laboratory of Developmental Pharmacology, at the National Institute of Child Health and Human Development (NICHHD) and the deputy chief of Laboratory of Developmental Pharmacology at NICHHD. Dr. Puga was the recipient of the Society of Toxicology Award in 1999 and of the University of Cincinnati College of Medicine Richard Akeson Award for Excellence in Teaching in 2002.

Andrew Renwick is an emeritus professor at the University of Southampton, having retired from his position as professor of biochemical pharmacology in September 2004. Dr. Renwick earned a B.Sc. degree in zoology and chemistry in 1967, a Ph.D. in biochemistry in 1971, and a D.Sc. (medicine) in pharmacology with toxicology in 1991 from the University of London. He was appointed lecturer in biochemistry at St. Mary's Hospital Medical School from 1969 until 1976. Dr. Renwick was senior lecturer in clinical pharmacology from 1976 to 1987 and was promoted to reader in clinical pharmacology in 1987 and professor of biochemical pharmacology in 1997.

Dr. Renwick's research interests focused on the absorption and metabolism of drugs and other foreign chemicals in humans following ingestion, inhalation, and dermal administration, in addition to species differences in the fate of chemicals in the body. His U.K. governmental advisory committee memberships have included the Medicines Commission, the Committee on Toxicity, and the Committee on Carcinogenicity of the Department of Health. In 2000, he was awarded an OBE (Officer [of the Order] of the British Empire) for services to U.K. Medicines Licensing Authority and Pharmacology and the Toxicology Forum George H. Scott Memorial Award in 2002. This award was presented in recognition of Dr. Renwick's efforts to promote the advancement and application of the science of toxicology with government, academics, and industry.

David Savitz is the Charles W. Bluhdorn Professor of Community and Preventive Medicine and the director of the Center of Excellence in Epidemiology, Biostatistics, and Disease Prevention at Mount Sinai School of Medicine. He earned a B.A. in 1975 from Brandeis University. In 1978, Dr. Savitz earned an M.S. from the Department of Preventive Medicine and then continued his education at the University of Pittsburgh, earning a Ph.D from the School of Public Health's Department of Epidemiology in 1982. He began his professional career as an assistant professor in the Department of Preventative Medicine at the University of Colorado School of Medicine. He joined the Department of Epidemiology at the University of North Carolina School of Public Health in 1985 and became chair of the department in 1996 and named Cary C. Boshamer Distinguished Professor in 2003. His research covers the areas of reproductive, environmental, and occupational epidemiology. Dr. Savitz is a member of many organizations and has served as president of the Society for Epidemiologic Research. Currently, Dr. Savitz is president of the Society for Pediatric and Perinatal Epidemiologic Research. He has authored over 200 peer-reviewed journal articles and is an editor of *Epidemiology*.

Allen Silverstone is professor of microbiology and immunology at SUNY–Upstate Medical University and adjunct professor of environmental medicine at the University of Rochester School of Medicine. Dr. Silverstone earned a B.A. from Reed College in 1965 and a Ph.D. from the Massachusetts Institute of Technology in 1970. His research interests include the cellular and molecular biology of how dioxins, estrogens, and estrogenic compounds affect the immune system. Having identified the particular target cell in T-cell development that is affected by dioxin, Dr. Silverstone's lab is now identifying the specific gene program activated by this agent in these cells. He was a member of the review panel and a consultant to the Science Advisory Board for EPA's reassessment of dioxin and related compounds.

Paul F. Terranova is professor of molecular and integrative physiology and obstetrics and gynecology. He is director of the Center for Reproductive Sciences at the University of Kansas Medical Center. He earned a B.S. in 1969 and an M.S. in 1971 in biology from McNeese State University and a Ph.D. from Louisiana State University in 1975. Dr. Terranova is an internationally recognized researcher in reproductive biology and has written or co-written more than 100 peer-reviewed original research papers, 19 chapters in books or symposium proceedings, and numerous articles in international scientific journals. He has served on numerous review panels of the National Institutes of Health, the National Science Foundation, and EPA. Dr. Terranova's research focuses on factors regulating follicular development and ovulation. Recently, he has found that an environmental contaminant, dioxin, prevents follicular rupture and he is assessing the endocrine and molecular mechanisms by which this blockage occurs. Dr. Terranova has also developed a mouse model of ovarian cancer and is determining which growth regulators are involved in the spontaneous transformation of the ovarian surface epithelial cells into a malignant phenotype.

Kimberly M. Thompson is an associate professor of risk analysis and decision science at the Harvard School of Public Health and Children's Hospital Boston. Professor Thompson recently joined the systems dynamics group at the Massachusetts Institute of Technology Sloan School of Management as a visitor. She earned a B.S. and an M.S. in chemical engineering from the Massachusetts Institute of Technology in 1988 and 1989, respectively, and an Sc.D. in environmental health from Harvard School of Health in 1995. Her research interests focus on issues related to developing and applying quantitative methods for risk assessment and risk management in addition to consideration of the public-policy implications associated with uncertainty and variability in risk characterization. Dr. Thompson is a member of many organizations and societies, including the Society for Risk Analysis and the International Society for Exposure Analysis. She was a Sigma Xi Distinguished Lecturer for 2003-2005 and has been the recipient of several honors, including recognition in 2003 by the Society of Toxicology for an outstanding published paper demonstrating an application of risk assessment with fellow colleagues and the 2004 Society for Risk Analysis Chauncey Starr Award.

Gary M. Williams is the director of environmental pathology and toxicology, head of the program on medicine, food, and chemical safety; and a professor of pathology at the New York Medical College since 1975. Dr. Williams earned a B.A. from Washington and Jefferson College in 1963 and an M.D. from the University of Pittsburgh School of Medicine in 1967. Following his residency at Massachusetts General Hospital, Dr. Williams began his career as assistant professor in the Department of Pathology at Temple University School of Medicine in 1971. He is board certified in pathology and toxicology. Dr. Williams' research focuses on mechanisms of chemical carcinogenesis and risk assessment. He has served on numerous working groups and committees of the NRC, EPA, International Agency for Research on Cancer, and World Health Organization. He has received many honors, including the Arnold J. Lehman and Enhancement of Animal Welfare Awards from the Society of Toxicology, and was elected fellow of the Royal College of Pathologists (U.K.).

Yiliang Zhu is a professor and director of the biostatistics Ph.D. program and the Center for Collaborative Research in the Department of Epidemiology and Biostatistics, College of Public Health, University of South Florida. Dr. Zhu earned a B.S. in computer science and applied mathematics from Shanghai University of Science and Technology (1982), an M.S. in statistics from Queen's University (1987), and a Ph.D. in statistics from the University of Toronto (1992). Dr. Zhu's research includes benchmark dose methods, dose response and PBPK modeling, and general methods in health risk assessment. Dr. Zhu has served on several EPA committees, including the Peer Review Committee on Neurobehavioral Dose-Response and Benchmark Method Guidance, Peer Review Committee on Benchmark Dose Software, Toxicological Review for 2-Methylnaphthalene, STAR Program Grant Review Committee for Global Change for Aquatic Ecosystems, and Peer Review Committee on Benchmark Doses Technical Guidance Document. Dr. Zhu is a member of the Advisory Committee on Organ Transplantation to the secretary of the U.S. Department of Health and Human Services.

B

EPA's 2005 Guidelines for Carcinogen Risk Assessment

"CARCINOGENIC TO HUMANS"

This descriptor indicates strong evidence of human carcinogenicity. It covers different combinations of evidence.

- This descriptor is appropriate when there is convincing epidemiologic evidence of a causal association between human exposure and cancer.
- Exceptionally, this descriptor may be equally appropriate with a lesser weight of epidemiologic evidence that is strengthened by other lines of evidence. It can be used when <u>all</u> of the following conditions are met: (a) there is strong evidence of an association between human exposure and either cancer or the key precursor events of the agent's mode of action but not enough for a causal association, <u>and</u> (b) there is extensive evidence of carcinogenicity in animals, <u>and</u> (c) the mode(s) of carcinogenic action and associated key precursor events have been identified in animals, <u>and</u> (d) there is strong evidence that the key precursor events that precede the cancer response in animals are anticipated to occur in humans and progress to tumors, based on available biological information. In this case, the narrative includes a summary of both the experimental and epidemiologic information on mode of action and also an indication of the relative weight that each source of information carries, e.g., based on human information, based on limited human and extensive animal experiments.

"LIKELY TO BE CARCINOGENIC TO HUMANS"

This descriptor is appropriate when the weight of the evidence is adequate to demonstrate carcinogenic potential to humans but does not reach the weight of evidence for the descriptor "Carcinogenic to Humans." Adequate evidence consistent with this descriptor covers a broad spectrum. As stated previously, the use of the term "likely" as a weight of evidence descriptor does not correspond to a quantifiable probability. The examples below are meant to represent the broad range of data combinations that are covered by this descriptor; they are illustrative and provide neither a checklist nor a limitation for the data that might support use of this descriptor. Moreover, additional information, e.g., on mode of action, might change the choice of descriptor for the illustrated examples. Supporting data for this descriptor may include

- an agent demonstrating a plausible (but not definitively causal) association between human exposure and cancer, in most cases with some supporting biological, experimental evidence, though not necessarily carcinogenicity data from animal experiments;
- an agent that has tested positive in animal experiments in more than one species, sex, strain, site, or exposure route, with or without evidence of carcinogenicity in humans;
- a positive tumor study that raises additional biological concerns beyond that of a statistically significant result, for example, a high degree of malignancy, or an early age at onset;
- a rare animal tumor response in a single experiment that is assumed to be relevant to humans; or
- a positive tumor study that is strengthened by other lines of evidence, for example, either plausible (but not definitively causal) association between human exposure and cancer <u>or</u> evidence that the agent or an important metabolite causes events generally known to be associated with tumor formation (such as DNA reactivity or effects on cell growth control) likely to be related to the tumor response in this case.

"SUGGESTIVE EVIDENCE OF CARCINOGENIC POTENTIAL"

This descriptor of the database is appropriate when the weight of evidence is suggestive of carcinogenicity; a concern for potential carcinogenic effects in humans is raised, but the data are judged not sufficient for a stronger conclusion. This descriptor covers a spectrum of evidence associated with varying levels of concern for carcinogenicity, ranging from a positive cancer result in the only study on an agent to a single positive cancer result in an extensive database that includes negative studies in other species. Depending on the extent of the database, additional studies may or may not provide further insights. Some examples include:

- a small, and possibly not statistically significant, increase in tumor incidence observed in a single animal or human study that does not reach the weight of evidence for the descriptor "Likely to Be Carcinogenic to Humans." The study generally would not be contradicted by other studies of equal quality in the same population group or experimental system (see discussions of *conflicting evidence* and *differing results*, below);
- a small increase in a tumor with a high background rate in that sex and strain, when there is some but insufficient evidence that the observed tumors may be due to intrinsic factors that cause background tumors and not due to the agent being assessed. (When there is a high background rate of a specific tumor in animals of a particular sex and strain, then there may be biological factors operating independently of the agent being assessed that could be responsible for the development of the observed tumors.) In this case, the reasons for determining that the tumors are not due to the agent are explained;
- evidence of a positive response in a study whose power, design, or conduct limits the ability to draw a confident conclusion (but does not make the study fatally flawed), but where the carcinogenic potential is strengthened by other lines of evidence (such as structure-activity relationships); or
- a statistically significant increase at one dose only, but no significant response at the other doses and no overall trend.

"INADEQUATE INFORMATION TO ASSESS CARCINOGENIC POTENTIAL"

This descriptor of the database is appropriate when available data are judged inadequate for applying one of the other descriptors. Additional studies generally would be expected to provide further insights. Some examples include:

- little or no pertinent information;
- conflicting evidence, that is, some studies provide evidence of carcinogenicity but other studies of equal quality in the same sex and strain are negative. *Differing results*, that is, positive results in some studies and negative results in one or more different experimental systems, do not constitute *conflicting evidence,* as the term is used here. Depending on the overall weight of evidence, differing results can be considered either suggestive evidence or likely evidence; or
- negative results that are not sufficiently robust for the descriptor, "Not Likely to Be Carcinogenic to Humans."

"NOT LIKELY TO BE CARCINOGENIC TO HUMANS"

This descriptor is appropriate when the available data are considered robust for deciding that there is no basis for human hazard concern. In some instances, there can be positive results in experimental animals when there is strong, consistent evidence that each mode of action in experimental animals does not operate in humans. In other cases, there can be convincing evidence in both humans and animals that the agent is not carcinogenic. The judgment may be based on data such as:

- animal evidence that demonstrates lack of carcinogenic effect in both sexes in well-designed and well-conducted studies in at least two appropriate animal species (in the absence of other animal or human data suggesting a potential for cancer effects),
- convincing and extensive experimental evidence showing that the only carcinogenic effects observed in animals are not relevant to humans,
- convincing evidence that carcinogenic effects are not likely by a particular exposure route (see Section 2.3), or
- convincing evidence that carcinogenic effects are not likely below a defined dose range.

A descriptor of "not likely" applies only to the circumstances supported by the data. For example, an agent may be "Not Likely to Be Carcinogenic" by one route but not necessarily by another. In those cases that have positive animal experiment(s) but the results are judged to be not relevant to humans, the narrative discusses why the results are not relevant.

MULTIPLE DESCRIPTORS

More than one descriptor can be used when an agent's effects differ by dose or exposure route. For example, an agent may be "Carcinogenic to Humans" by one exposure route but "Not Likely to Be Carcinogenic" by a route by which it is not absorbed. Also, an agent could be "Likely to Be Carcinogenic" above a specified dose but "Not Likely to Be Carcinogenic" below that dose because a key event in tumor formation does not occur below that dose.